Israel Population Growth

from Genesis to Exodus

Stanley R. Stasko

authorHOUSE®

AuthorHouse™
1663 Liberty Drive
Bloomington, IN 47403
www.authorhouse.com
Phone: 1-800-839-8640

First published by AuthorHouse 5/6/2009

ISBN: 978-1-4389-6629-8 (sc)

Printed in the United States of America
Bloomington, Indiana

This book is printed on acid-free paper.

This book is dedicated to Stanley R. Stasko spiritual experience on May 12, 2003 and May 13, 2003.

Thank you spiritual voice for answering a significant theological question:

Protestants are wrong,
once in faith,
not always in faith.

The author gave more details about this spiritual experience in his application for acceptance as a seminarian for the Diocese of Madison, Wisconsin, approximately August 2004.

Introduction:

In the study of Sacred Scripture some people believe that the Exodus of Israel from Egypt is myth and other people believe the Exodus of Israel from Egypt is a historic event. In 1995 the author entered Sacred Heart Major Seminary in Detroit, Michigan in priestly formation. In 1996 he received a Bachelor of Philosophy from Sacred Heart Major Seminary and left priestly formation. In the fall of 1996 the author enrolled in a graduate course at Sacred Heart Major Seminary entitled Method & Pentateuch (course #SS-521). During the course the instructor implied that the Exodus of Israel from Egypt had to be myth because the birth rate of children per woman during Israel stay in Egypt had to be so high (10, 11, 12 children per woman) and this is a unreasonable high birth rate per woman.

This book will address the question "Can anyone make the claim that the Exodus of Israel from Egypt is myth based on the singular argument that the birth rate of children per woman during Israel stay in Egypt is so high (10, 11, 12 children per woman) and therefore is unreasonable."

The answer begins by asking the question what birth rate of children per woman during Israel stay in Egypt is required to obtain Israel population growth described in Genesis to Exodus. The author examines this question utilizing two methods: a machine design analysis of Israel population growth from Genesis to Exodus and a statistical numerical analysis of Israel population growth from Genesis to Exodus.

The machine design analysis requires a birth rate of 3.366 to 4.252 children per woman to obtain Israel population growth from Genesis to Exodus. The statistical numerical analysis requires a birth rate of 3.540 to 4.167 children per woman of to obtain Israel population growth from Genesis to Exodus.

The claim that the Exodus of Israel from Egypt is myth based on the singular argument that the birth rate required of children per woman during Israel stay in Egypt is so high (10, 11, 12 children per woman) is mathematically invalid.

Machine Design Analysis

Machine design analysis to calculate the birth rate of children per woman during Israel stay in Egypt to obtain Israel population growth from Genesis to Exodus might be unfamiliar to the reader. Let us proceed by using a thought experiment.

The author lives in Southfield, Michigan. Near the corner of Eleven Mile Road and Lahser Road is an United States Postal Office. Inside the United States Postal Office is at least one automated postal center. Imagine your employer asking you to design an automated postal center for that United States Postal Center.

A machine designer might begin by asking how many machines need to be manufactured and how many customers are expected to use the automated postal center. Israel population growth from Genesis to Exodus is a singular event in the history of Israel; therefore, the quantity of machines needed to be manufactured is one and the expected number of customers is one.

The machine designer might next ask for a set of specifications in the machine design analysis. Sacred Scripture provides us with limited information.

Exodus 12: 40

The time the Israelites had stayed in Egypt was four hundred and thirty years.

All Sacred Scripture references are from the Saint Joseph Edition of The New American Bible.

Genesis 46: 27

Together with Joseph's sons who were born to him in Egypt – two persons – all the people comprising Jacob's family who had come to Egypt amounted to seventy persons in all.

Genesis 46: 26-27

Jacob's people who migrated to Egypt – his direct descendants, not counting the wives of Jacob's sons – numbered sixty-six persons in all. Together with Joseph's sons who were born to him in Egypt – two persons – all the people comprising Jacob's family who had come to Egypt amounted to seventy persons in all.

Numbers 1: 1-3

In the year following that of the Israelites' departure from the land of Egypt, on the first day of the second month, the Lord said to Moses in the meeting tent in the desert of Sinai: "Take a census of the whole community of the Israelites, by clans and ancestral houses, registering each male individually. You and Aaron shall enroll in companies all the men in Israel of twenty years or more who are fit for military service.

Numbers 1: 44 – 46

It was these who were registered, each according to his ancestral house, in the census taken by Moses and Aaron and the twelve princes of Israel. The total number of the Israelites of twenty years or more who were fit for military service, registered by ancestral houses, was six hundred and three thousand five hundred and fifty.

Numbers 1: 47-49

The Levites, however, were not registered by ancestral tribe with the others. For the Lord had told Moses, "The tribe of Levi alone you shall not enroll nor include in the census alone with the other Israelites.

Numbers 3: 39

the total number of male Levites a month old or more whom Moses had registered by clans in keeping with the Lord's command, was twenty-two thousand.

Psalm 90: 8-10

You have kept our faults before you, our hidden sins exposed to your sight. Our life ebbs away under your wrath; our years end like a sigh. Seventy is the sum of our years, or eighty, if we are strong; Most of them are sorrow and toil; they pass quickly, we are all but gone.

Genesis 46:27 states that Jacob's family amounted to seventy persons in all. Genesis 46: 26-27 implies Jacob's people amounted to sixty-six persons. The starting population of Israel population growth from Genesis to Exodus is seventy or sixty-six persons.

Starting Population

70	66

Genesis 46: 27 states that Jacob's family amounted to seventy persons in all. Seventy persons being men, women, and children (M/W/C). Genesis 46: 26-27 states that Jacob's people who migrated to Egypt – his direct descendants, not counting the wives of Jacob's sons – numbered sixty-six persons in all. Sixty-six persons being males only (M no W) and we need to account for females. Numbers 1: 1-3 and Numbers 1: 44-46 total number of adult men only (M only) was 603,550 and we need to account for adult women, male children, and female children.

Sacred Scripture does not provide a year by year U.S. Census style information; therefore, Israel starting population is one of three possibilities (M/W/C), (M no W), or (M only).

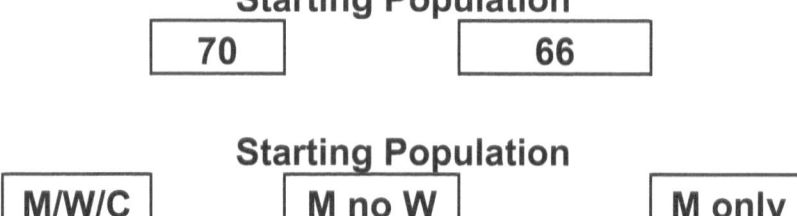

Starting Population

70	66

Starting Population

M/W/C	M no W	M only

Israel starting population varies significantly between 70 (M/W/C), 70 (M no W), and 70 (M only). Later in the book machine design analysis 001 will determine that a birth rate of 3.584 children per woman is required to obtain Israel population growth from Genesis to Exodus.

70 (M/W/C) = Starting population of 70 persons total

70 (M no W) = Starting population of (70 * 2) equals 140 persons total

70 (M only) = Starting population of (70 * 2) + (70 * 3.584) equals 391 persons total

Numbers 1: 1-3 implies Israel ending population at 603,550. Numbers 1: 47-49 states that the tribe of Levi was not included in the census of the Israelites. Numbers 3:39 states the total number male Levites was 22,000. Therefore, the ending population of Israel population growth from Genesis to Exodus is 603,550 or 625,550 persons.

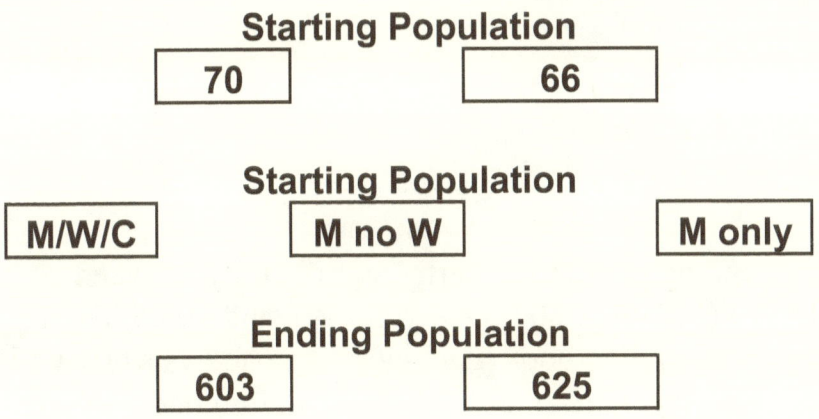

Israel ending population can have different interpretations similar to Israel starting population. Israel ending population could be men, women, and children (M/W/C), males only (M no W) and we need to account for females, or adult men only (M only) and we need to account for adult women, male children, and female children.

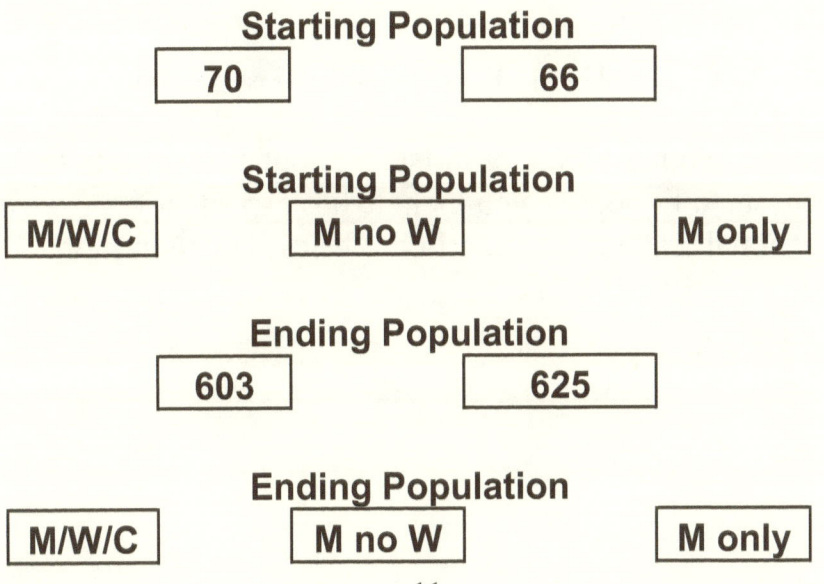

Israel ending population varies significantly between 603,550 (M/W/C), 603,550 (M no W), and 603,550 (M only). Later in the book machine design analysis 001 will determine that a birth rate of 3.584 children per woman is required to obtain Israel population growth from Genesis to Exodus.

603,550 (M/W/C) = Ending population of 603,550 persons total

603,550 (M no W) = Ending population of (603,550 * 2) equals 1,207,100 persons total

603,550 (M only) = Ending population of (603,550 * 2) + (603,550 * 3.584) equals 3,370,223 persons total

There are other variables in Israel population growth from Genesis to Exodus which include life expectancy, fractional, and infant mortality to be explained in more detail later in the book.

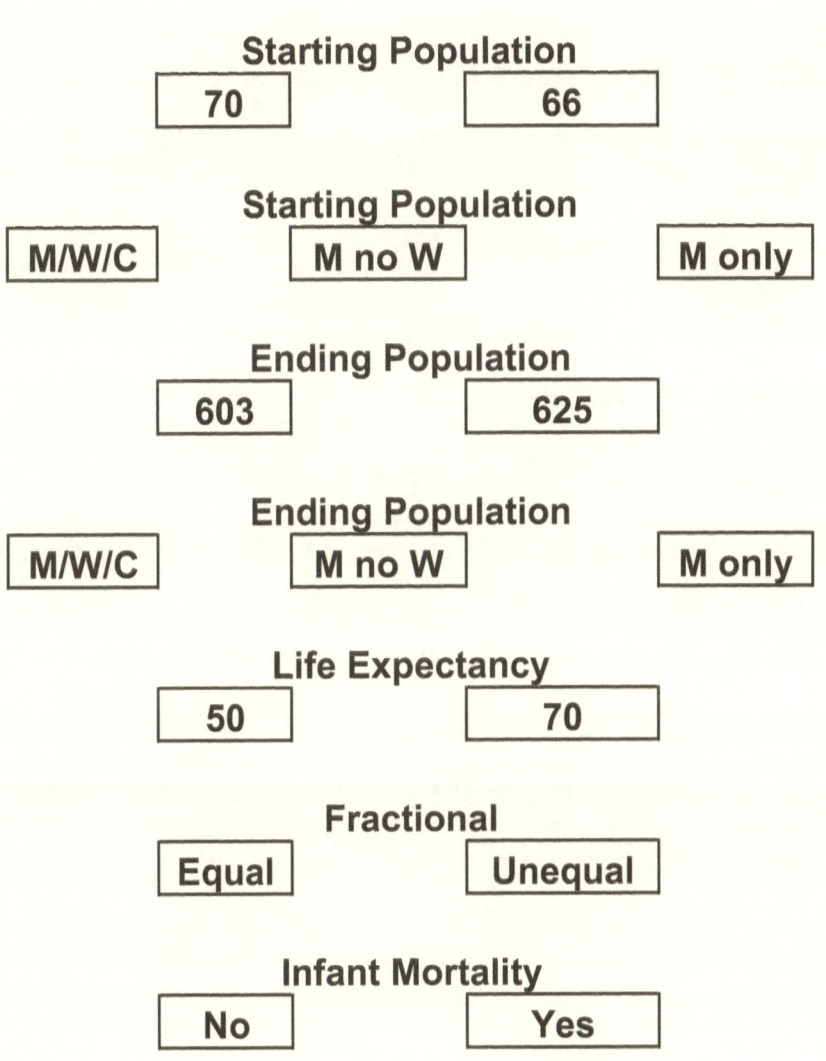

Starting Population

| 70 | 66 |

Starting Population

| M/W/C | M no W | M only |

Ending Population

| 603 | 625 |

Ending Population

| M/W/C | M no W | M only |

Life Expectancy

| 50 | 70 |

Fractional

| Equal | Unequal |

Infant Mortality

| No | Yes |

Earlier the author stated that a machine designer might begin by asking how many machines need to be manufactured and how many customers are expected to use the automated postal center. Israel population growth from Genesis to Exodus is a singular event in the history of Israel; therefore, the quantity of machines needed to be manufactured is one and the expected number of customers is one. Since the there are 288 total number of possible machine design analysis combinations (2 * 3 * 2 * 3 * 2 * 2 *2 equals 288), and since no reasonable machine designer would analyze 288 machine design combinations for one machine to be used by one customer the author suggests the following 18 machine design combinations as adequate for analyzing Israel population growth from Genesis to Exodus.

Design	Starting Population	Ending Population	Life Expectancy	Fractional	Infant Mortality	Birth Rate Required
001	70 M/W/C	603 M/W/C	50	Equal	No	
002	66 M/W/C	603 M/W/C	50	Equal	No	
009	70 M no W	625 M/W/C	50	Equal	No	
046	66 M no W	625 M/W/C	70	Equal	No	
113	70 M only	603 M/W/C	70	Unequal	No	
114	66 M only	603 M/W/C	70	Unequal	No	
235	70 M/W/C	625 M no W	50	Unequal	Yes	
236	66 M/W/C	625 M no W	50	Unequal	Yes	
159	70 M no W	603 M no W	50	Equal	Yes	
196	66 M no W	603 M no W	70	Equal	Yes	
203	70 M only	625 M no W	70	Equal	Yes	
204	66 M only	625 M no W	70	Equal	Yes	
097	70 M/W/C	603 M only	50	Unequal	No	
098	66 M/W/C	603 M only	50	Unequal	No	
105	70 M no W	625 M only	50	Unequal	No	
142	66 M no W	625 M only	70	Unequal	No	
065	70 M only	603 M only	70	Equal	No	
066	66 M only	603 M only	70	Equal	No	

Machine Design Analysis 001

Starting Population

70	

Starting Population

M/W/C		

Ending Population

603	

Ending Population

M/W/C		

Life Expectancy

50	

Fractional

Equal	

Infant Mortality

No	

For the starting population (Genesis 46: 27) assume the word "men" means adult men, adult women, and children (70 M/W/C); therefore, the total starting population would be 70 persons.

Assume the time the Israelites had stayed in Egypt was four hundred and thirty years (Exodus 12:40).

For the ending population (Numbers 1: 1-3 and Numbers 1: 44-46) assume the word "men" means adult men, adult women, and children (603 M/W/C); therefore, the total ending population would be 603,550 persons.

Assume a human life expectancy of 50 years.

Using numerical analysis to calculate the birth rate required for a total starting population of 70 persons to increase to a total ending population of 603,550 persons in 430 years results in a birth rate of 3.584 children.

Numerical analysis 1

Birth rate per (2) persons per (20) years — 3.584
Birth rate per (1) person per (20) years — 1.792
Birth rate per (1) person per (5) years — 0.448

Total starting population of 70 persons.
Total ending population of 603,550 persons
Israel stayed in Egypt 430 years.

0 to 5 Yrs	5 to 10 Yrs	10 to 15 Yrs	15 to 20 Yrs	20 to 25 Yrs	25 to 30 Yrs	30 to 35 Yrs	35 to 40 Yrs	40 to 45 Yrs	45 to 50 Yrs	Total Population	Years
7.0	7.0	7.0	7.0	7.0	7.0	7.0	7.0	7.0	7.0	70.0	0
12.5	7.0	7.0	7.0	7.0	7.0	7.0	7.0	7.0	7.0	75.5	5
12.5	12.5	7.0	7.0	7.0	7.0	7.0	7.0	7.0	7.0	81.1	10
12.5	12.5	12.5	7.0	7.0	7.0	7.0	7.0	7.0	7.0	86.6	15
12.5	12.5	12.5	12.5	7.0	7.0	7.0	7.0	7.0	7.0	92.2	20
15.0	12.5	12.5	12.5	12.5	7.0	7.0	7.0	7.0	7.0	100.2	25
17.5	15.0	12.5	12.5	12.5	12.5	7.0	7.0	7.0	7.0	110.7	30
20.0	17.5	15.0	12.5	12.5	12.5	12.5	7.0	7.0	7.0	123.7	35
22.5	20.0	17.5	15.0	12.5	12.5	12.5	12.5	7.0	7.0	139.2	40
23.6	22.5	20.0	17.5	15.0	12.5	12.5	12.5	12.5	7.0	155.8	45
25.8	23.6	22.5	20.0	17.5	15.0	12.5	12.5	12.5	12.5	174.6	50
29.2	25.8	23.6	22.5	20.0	17.5	15.0	12.5	12.5	12.5	191.2	55
33.6	29.2	25.8	23.6	22.5	20.0	17.5	15.0	12.5	12.5	212.3	60
37.4	33.6	29.2	25.8	23.6	22.5	20.0	17.5	15.0	12.5	237.2	65
41.2	37.4	33.6	29.2	25.8	23.6	22.5	20.0	17.5	15.0	265.8	70
45.3	41.2	37.4	33.6	29.2	25.8	23.6	22.5	20.0	17.5	296.0	75
50.3	45.3	41.2	37.4	33.6	29.2	25.8	23.6	22.5	20.0	328.8	80
56.5	50.3	45.3	41.2	37.4	33.6	29.2	25.8	23.6	22.5	365.2	85
63.3	56.5	50.3	45.3	41.2	37.4	33.6	29.2	25.8	23.6	406.1	90
70.6	63.3	56.5	50.3	45.3	41.2	37.4	33.6	29.2	25.8	453.0	95
78.0	70.6	63.3	56.5	50.3	45.3	41.2	37.4	33.6	29.2	505.2	100
86.5	78.0	70.6	63.3	56.5	50.3	45.3	41.2	37.4	33.6	562.6	105
96.5	86.5	78.0	70.6	63.3	56.5	50.3	45.3	41.2	37.4	625.5	110
107.8	96.5	86.5	78.0	70.6	63.3	56.5	50.3	45.3	41.2	695.8	115
120.2	107.8	96.5	86.5	78.0	70.6	63.3	56.5	50.3	45.3	774.9	120
133.7	120.2	107.8	96.5	86.5	78.0	70.6	63.3	56.5	50.3	863.3	125
148.5	133.7	120.2	107.8	96.5	86.5	78.0	70.6	63.3	56.5	961.6	130
165.2	148.5	133.7	120.2	107.8	96.5	86.5	78.0	70.6	63.3	1,070.3	135
184.1	165.2	148.5	133.7	120.2	107.8	96.5	86.5	78.0	70.6	1,191.1	140
205.3	184.1	165.2	148.5	133.7	120.2	107.8	96.5	86.5	78.0	1,325.8	145
228.6	205.3	184.1	165.2	148.5	133.7	120.2	107.8	96.5	86.5	1,476.4	150
254.3	228.6	205.3	184.1	165.2	148.5	133.7	120.2	107.8	96.5	1,644.2	155
282.9	254.3	228.6	205.3	184.1	165.2	148.5	133.7	120.2	107.8	1,830.6	160
315.0	282.9	254.3	228.6	205.3	184.1	165.2	148.5	133.7	120.2	2,037.9	165
350.9	315.0	282.9	254.3	228.6	205.3	184.1	165.2	148.5	133.7	2,268.5	170
390.8	350.9	315.0	282.9	254.3	228.6	205.3	184.1	165.2	148.5	2,525.6	175
435.0	390.8	350.9	315.0	282.9	254.3	228.6	205.3	184.1	165.2	2,812.1	180
484.2	435.0	390.8	350.9	315.0	282.9	254.3	228.6	205.3	184.1	3,131.1	185
539.0	484.2	435.0	390.8	350.9	315.0	282.9	254.3	228.6	205.3	3,486.0	190
600.1	539.0	484.2	435.0	390.8	350.9	315.0	282.9	254.3	228.6	3,880.8	195
668.3	600.1	539.0	484.2	435.0	390.8	350.9	315.0	282.9	254.3	4,320.5	200
744.1	668.3	600.1	539.0	484.2	435.0	390.8	350.9	315.0	282.9	4,810.3	205

0 to 5 Yrs	5 to 10 Yrs	10 to 15 Yrs	15 to 20 Yrs	20 to 25 Yrs	25 to 30 Yrs	30 to 35 Yrs	35 to 40 Yrs	40 to 45 Yrs	45 to 50 Yrs	Total Population	Years
828.4	744.1	668.3	600.1	539.0	484.2	435.0	390.8	350.9	315.0	5,355.7	210
922.2	828.4	744.1	668.3	600.1	539.0	484.2	435.0	390.8	350.9	5,962.9	215
1,026.6	922.2	828.4	744.1	668.3	600.1	539.0	484.2	435.0	390.8	6,638.7	220
1,143.1	1,026.6	922.2	828.4	744.1	668.3	600.1	539.0	484.2	435.0	7,390.9	225
1,272.7	1,143.1	1,026.6	922.2	828.4	744.1	668.3	600.1	539.0	484.2	8,228.6	230
1,417.0	1,272.7	1,143.1	1,026.6	922.2	828.4	744.1	668.3	600.1	539.0	9,161.3	235
1,577.5	1,417.0	1,272.7	1,143.1	1,026.6	922.2	828.4	744.1	668.3	600.1	10,199.9	240
1,756.3	1,577.5	1,417.0	1,272.7	1,143.1	1,026.6	922.2	828.4	744.1	668.3	11,356.0	245
1,955.3	1,756.3	1,577.5	1,417.0	1,272.7	1,143.1	1,026.6	922.2	828.4	744.1	12,643.0	250
2,177.0	1,955.3	1,756.3	1,577.5	1,417.0	1,272.7	1,143.1	1,026.6	922.2	828.4	14,075.9	255
2,423.8	2,177.0	1,955.3	1,756.3	1,577.5	1,417.0	1,272.7	1,143.1	1,026.6	922.2	15,671.4	260
2,698.5	2,423.8	2,177.0	1,955.3	1,756.3	1,577.5	1,417.0	1,272.7	1,143.1	1,026.6	17,447.7	265
3,004.3	2,698.5	2,423.8	2,177.0	1,955.3	1,756.3	1,577.5	1,417.0	1,272.7	1,143.1	19,425.4	270
3,344.8	3,004.3	2,698.5	2,423.8	2,177.0	1,955.3	1,756.3	1,577.5	1,417.0	1,272.7	21,627.1	275
3,723.9	3,344.8	3,004.3	2,698.5	2,423.8	2,177.0	1,955.3	1,756.3	1,577.5	1,417.0	24,078.4	280
4,146.1	3,723.9	3,344.8	3,004.3	2,698.5	2,423.8	2,177.0	1,955.3	1,756.3	1,577.5	26,807.4	285
4,616.0	4,146.1	3,723.9	3,344.8	3,004.3	2,698.5	2,423.8	2,177.0	1,955.3	1,756.3	29,845.9	290
5,139.2	4,616.0	4,146.1	3,723.9	3,344.8	3,004.3	2,698.5	2,423.8	2,177.0	1,955.3	33,228.9	295
5,721.6	5,139.2	4,616.0	4,146.1	3,723.9	3,344.8	3,004.3	2,698.5	2,423.8	2,177.0	36,995.2	300
6,370.2	5,721.6	5,139.2	4,616.0	4,146.1	3,723.9	3,344.8	3,004.3	2,698.5	2,423.8	41,188.4	305
7,092.2	6,370.2	5,721.6	5,139.2	4,616.0	4,146.1	3,723.9	3,344.8	3,004.3	2,698.5	45,856.8	310
7,896.1	7,092.2	6,370.2	5,721.6	5,139.2	4,616.0	4,146.1	3,723.9	3,344.8	3,004.3	51,054.3	315
8,791.1	7,896.1	7,092.2	6,370.2	5,721.6	5,139.2	4,616.0	4,146.1	3,723.9	3,344.8	56,841.1	320
9,787.4	8,791.1	7,896.1	7,092.2	6,370.2	5,721.6	5,139.2	4,616.0	4,146.1	3,723.9	63,283.7	325
10,896.8	9,787.4	8,791.1	7,896.1	7,092.2	6,370.2	5,721.6	5,139.2	4,616.0	4,146.1	70,456.6	330
12,131.9	10,896.8	9,787.4	8,791.1	7,896.1	7,092.2	6,370.2	5,721.6	5,139.2	4,616.0	78,442.4	335
13,507.0	12,131.9	10,896.8	9,787.4	8,791.1	7,896.1	7,092.2	6,370.2	5,721.6	5,139.2	87,333.4	340
15,037.9	13,507.0	12,131.9	10,896.8	9,787.4	8,791.1	7,896.1	7,092.2	6,370.2	5,721.6	97,232.1	345
16,742.4	15,037.9	13,507.0	12,131.9	10,896.8	9,787.4	8,791.1	7,896.1	7,092.2	6,370.2	108,252.8	350
18,640.0	16,742.4	15,037.9	13,507.0	12,131.9	10,896.8	9,787.4	8,791.1	7,896.1	7,092.2	120,522.7	355
20,752.7	18,640.0	16,742.4	15,037.9	13,507.9	12,131.9	10,896.8	9,787.4	8,791.1	7,896.1	134,183.2	360
23,104.9	20,752.7	18,640.0	16,742.4	15,037.9	13,507.9	12,131.9	10,896.8	9,787.4	8,791.1	149,392.1	365
25,723.8	23,104.9	20,752.7	18,640.0	16,742.4	15,037.9	13,507.9	12,131.9	10,896.8	9,787.4	166,324.8	370
28,639.4	25,723.8	23,104.9	20,752.7	18,640.0	16,742.4	15,037.9	13,507.9	12,131.9	10,896.8	185,176.7	375
31,885.5	28,639.4	25,723.8	23,104.9	20,752.7	18,640.0	16,742.4	15,037.9	13,507.9	12,131.9	206,165.5	380
35,499.5	31,885.5	28,639.4	25,723.8	23,104.9	20,752.7	18,640.0	16,742.4	15,037.9	13,507.9	229,533.1	385
39,523.2	35,499.5	31,885.5	28,639.4	25,723.8	23,104.9	20,752.7	18,640.0	16,742.4	15,037.9	255,549.4	390
44,002.9	39,523.2	35,499.5	31,885.5	28,639.4	25,723.8	23,104.9	20,752.7	18,640.0	16,742.4	284,514.4	395
48,990.4	44,002.9	39,523.2	35,499.5	31,885.5	28,639.4	25,723.8	23,104.9	20,752.7	18,640.0	316,762.5	400
54,543.2	48,990.4	44,002.9	39,523.2	35,499.5	31,885.5	28,639.4	25,723.8	23,104.9	20,752.7	352,665.6	405
60,725.4	54,543.2	48,990.4	44,002.9	39,523.2	35,499.5	31,885.5	28,639.4	25,723.8	23,104.9	392,638.4	410
67,608.2	60,725.4	54,543.2	48,990.4	44,002.9	39,523.2	35,499.5	31,885.5	28,639.4	25,723.8	437,141.5	415
75,271.2	67,608.2	60,725.4	54,543.2	48,990.4	44,002.9	39,523.2	35,499.5	31,885.5	28,639.4	486,689.0	420
83,802.8	75,271.2	67,608.2	60,725.4	54,543.2	48,990.4	44,002.9	39,523.2	35,499.5	31,885.5	541,852.4	425
93,301.3	83,802.8	75,271.2	67,608.2	60,725.4	54,543.2	48,990.4	44,002.9	39,523.2	35,499.5	603,268.2	430

Design	Starting Population	Ending Population	Life Expectancy	Fractional	Infant Mortality	Birth Rate Required
001	70	603,550	50	Equal	No	3.584
002	66 M/W/C	603 M/W/C	50	Equal	No	
009	70 M no W	625 M/W/C	50	Equal	No	
046	66 M no W	625 M/W/C	70	Equal	No	
113	70 M only	603 M/W/C	70	Unequal	No	
114	66 M only	603 M/W/C	70	Unequal	No	
235	70 M/W/C	625 M no W	50	Unequal	Yes	
236	66 M/W/C	625 M no W	50	Unequal	Yes	
159	70 M no W	603 M no W	50	Equal	Yes	
196	66 M no W	603 M no W	70	Equal	Yes	
203	70 M only	625 M no W	70	Equal	Yes	
204	66 M only	625 M no W	70	Equal	Yes	
097	70 M/W/C	603 M only	50	Unequal	No	
098	66 M/W/C	603 M only	50	Unequal	No	
105	70 M no W	625 M only	50	Unequal	No	
142	66 M no W	625 M only	70	Unequal	No	
065	70 M only	603 M only	70	Equal	No	
066	66 M only	603 M only	70	Equal	No	

Question: Why does the author use a life expectancy of 50 years in his analysis of Israel population growth from Genesis to Exodus?

Answer: If the author used 75, 80, 85 or 90 year life expectancy in his analysis of Israel population growth from Genesis to Exodus then people who claim Sacred Scripture is myth would argue that the author is using unrealistic life expectancies in his numerical analysis. Later the author will also use a life expectancy of 70 years (see Psalm 90: 8-10). This book uses 50 and 70 year life expectancies as the two possibilities when analyzing Israel population growth from Genesis to Exodus.

Machine Design Analysis 002

Starting Population

| | 66 |

Starting Population

| M/W/C | | |

Ending Population

| 603 | |

Ending Population

| M/W/C | | |

Life Expectancy

| 50 | |

Fractional

| Equal | |

Infant Mortality

| No | |

For the starting population (Genesis 46: 26-27) assume the word "men" means adult men, adult women, and children (66 M/W/C); therefore, the total starting population would be 66 persons.

Assume the time the Israelites had stayed in Egypt was four hundred and thirty years (Exodus 12:40).

For the ending population (Numbers 1: 1-3 and Numbers 1: 44-46) assume the word "men" means adult men, adult women, and children (603 M/W/C); therefore, the total ending population would be 603,550 persons.

Assume a human life expectancy of 50 years. Using numerical analysis to calculate the birth rate required for a total starting population of 66 persons to increase to a total ending population of 603,550 persons in 430 years results in a birth rate of 3.598 children.

Numerical analysis 2

Birth rate per (2) persons per (20) years	3.598
Birth rate per (1) person per (20) years	1.799
Birth rate per (1) person per (5) years	0.450

Total starting population of 66 persons.
Total ending population of 603,550 persons
Israel stayed in Egypt 430 years.

0 to 5 Yrs	5 to 10 Yrs	10 to 15 Yrs	15 to 20 Yrs	20 to 25 Yrs	25 to 30 Yrs	30 to 35 Yrs	35 to 40 Yrs	40 to 45 Yrs	45 to 50 Yrs	Total Population	Years
6.6	6.6	6.6	6.6	6.6	6.6	6.6	6.6	6.6	6.6	66.0	0
11.9	6.6	6.6	6.6	6.6	6.6	6.6	6.6	6.6	6.6	71.3	5
11.9	11.9	6.6	6.6	6.6	6.6	6.6	6.6	6.6	6.6	76.5	10
11.9	11.9	11.9	6.6	6.6	6.6	6.6	6.6	6.6	6.6	81.8	15
11.9	11.9	11.9	11.9	6.6	6.6	6.6	6.6	6.6	6.6	87.1	20
14.2	11.9	11.9	11.9	11.9	6.6	6.6	6.6	6.6	6.6	94.7	25
16.6	14.2	11.9	11.9	11.9	11.9	6.6	6.6	6.6	6.6	104.8	30
19.0	16.6	14.2	11.9	11.9	11.9	11.9	6.6	6.6	6.6	117.1	35
21.4	19.0	16.6	14.2	11.9	11.9	11.9	11.9	6.6	6.6	131.9	40
22.4	21.4	19.0	16.6	14.2	11.9	11.9	11.9	11.9	6.6	147.7	45
24.6	22.4	21.4	19.0	16.6	14.2	11.9	11.9	11.9	11.9	165.7	50
27.8	24.6	22.4	21.4	19.0	16.6	14.2	11.9	11.9	11.9	181.6	55
32.0	27.8	24.6	22.4	21.4	19.0	16.6	14.2	11.9	11.9	201.7	60
35.7	32.0	27.8	24.6	22.4	21.4	19.0	16.6	14.2	11.9	225.6	65
39.3	35.7	32.0	27.8	24.6	22.4	21.4	19.0	16.6	14.2	253.0	70
43.2	39.3	35.7	32.0	27.8	24.6	22.4	21.4	19.0	16.6	282.0	75
48.0	43.2	39.3	35.7	32.0	27.8	24.6	22.4	21.4	19.0	313.4	80
54.0	48.0	43.2	39.3	35.7	32.0	27.8	24.6	22.4	21.4	348.4	85
60.6	54.0	48.0	43.2	39.3	35.7	32.0	27.8	24.6	22.4	387.6	90
67.6	60.6	54.0	48.0	43.2	39.3	35.7	32.0	27.8	24.6	432.8	95
74.8	67.6	60.6	54.0	48.0	43.2	39.3	35.7	32.0	27.8	483.0	100
83.0	74.8	67.6	60.6	54.0	48.0	43.2	39.3	35.7	32.0	538.2	105
92.6	83.0	74.8	67.6	60.6	54.0	48.0	43.2	39.3	35.7	598.7	110
103.5	92.6	83.0	74.8	67.6	60.6	54.0	48.0	43.2	39.3	666.6	115
115.6	103.5	92.6	83.0	74.8	67.6	60.6	54.0	48.0	43.2	742.9	120
128.6	115.6	103.5	92.6	83.0	74.8	67.6	60.6	54.0	48.0	828.2	125
143.0	128.6	115.6	103.5	92.6	83.0	74.8	67.6	60.6	54.0	923.2	130
159.2	143.0	128.6	115.6	103.5	92.6	83.0	74.8	67.6	60.6	1,028.3	135
177.5	159.2	143.0	128.6	115.6	103.5	92.6	83.0	74.8	67.6	1,145.2	140
198.0	177.5	159.2	143.0	128.6	115.6	103.5	92.6	83.0	74.8	1,275.7	145
220.7	198.0	177.5	159.2	143.0	128.6	115.6	103.5	92.6	83.0	1,421.6	150
245.7	220.7	198.0	177.5	159.2	143.0	128.6	115.6	103.5	92.6	1,584.3	155
273.6	245.7	220.7	198.0	177.5	159.2	143.0	128.6	115.6	103.5	1,765.3	160
304.8	273.6	245.7	220.7	198.0	177.5	159.2	143.0	128.6	115.6	1,966.5	165
339.7	304.8	273.6	245.7	220.7	198.0	177.5	159.2	143.0	128.6	2,190.6	170
378.6	339.7	304.8	273.6	245.7	220.7	198.0	177.5	159.2	143.0	2,440.7	175
421.8	378.6	339.7	304.8	273.6	245.7	220.7	198.0	177.5	159.2	2,719.5	180
469.8	421.8	378.6	339.7	304.8	273.6	245.7	220.7	198.0	177.5	3,030.2	185
523.4	469.8	421.8	378.6	339.7	304.8	273.6	245.7	220.7	198.0	3,376.1	190
583.2	523.4	469.8	421.8	378.6	339.7	304.8	273.6	245.7	220.7	3,761.3	195
649.9	583.2	523.4	469.8	421.8	378.6	339.7	304.8	273.6	245.7	4,190.5	200
724.1	649.9	583.2	523.4	469.8	421.8	378.6	339.7	304.8	273.6	4,668.9	205

0 to 5 Yrs	5 to 10 Yrs	10 to 15 Yrs	15 to 20 Yrs	20 to 25 Yrs	25 to 30 Yrs	30 to 35 Yrs	35 to 40 Yrs	40 to 45 Yrs	45 to 50 Yrs	Total Population	Years
806.7	724.1	649.9	583.2	523.4	469.8	421.8	378.6	339.7	304.8	5,202.1	210
898.7	806.7	724.1	649.9	583.2	523.4	469.8	421.8	378.6	339.7	5,796.0	215
1,001.3	898.7	806.7	724.1	649.9	583.2	523.4	469.8	421.8	378.6	6,457.6	220
1,115.6	1,001.3	898.7	806.7	724.1	649.9	583.2	523.4	469.8	421.8	7,194.6	225
1,243.1	1,115.6	1,001.3	898.7	806.7	724.1	649.9	583.2	523.4	469.8	8,015.8	230
1,385.0	1,243.1	1,115.6	1,001.3	898.7	806.7	724.1	649.9	583.2	523.4	8,930.9	235
1,543.0	1,385.0	1,243.1	1,115.6	1,001.3	898.7	806.7	724.1	649.9	583.2	9,950.5	240
1,719.1	1,543.0	1,385.0	1,243.1	1,115.6	1,001.3	898.7	806.7	724.1	649.9	11,086.4	245
1,915.3	1,719.1	1,543.0	1,385.0	1,243.1	1,115.6	1,001.3	898.7	806.7	724.1	12,351.9	250
2,134.0	1,915.3	1,719.1	1,543.0	1,385.0	1,243.1	1,115.6	1,001.3	898.7	806.7	13,761.8	255
2,377.7	2,134.0	1,915.3	1,719.1	1,543.0	1,385.0	1,243.1	1,115.6	1,001.3	898.7	15,332.8	260
2,649.1	2,377.7	2,134.0	1,915.3	1,719.1	1,543.0	1,385.0	1,243.1	1,115.6	1,001.3	17,083.2	265
2,951.4	2,649.1	2,377.7	2,134.0	1,915.3	1,719.1	1,543.0	1,385.0	1,243.1	1,115.6	19,033.3	270
3,288.3	2,951.4	2,649.1	2,377.7	2,134.0	1,915.3	1,719.1	1,543.0	1,385.0	1,243.1	21,206.0	275
3,663.7	3,288.3	2,951.4	2,649.1	2,377.7	2,134.0	1,915.3	1,719.1	1,543.0	1,385.0	23,626.7	280
4,082.0	3,663.7	3,288.3	2,951.4	2,649.1	2,377.7	2,134.0	1,915.3	1,719.1	1,543.0	26,323.7	285
4,548.0	4,082.0	3,663.7	3,288.3	2,951.4	2,649.1	2,377.7	2,134.0	1,915.3	1,719.1	29,328.7	290
5,067.1	4,548.0	4,082.0	3,663.7	3,288.3	2,951.4	2,649.1	2,377.7	2,134.0	1,915.3	32,676.7	295
5,645.5	5,067.1	4,548.0	4,082.0	3,663.7	3,288.3	2,951.4	2,649.1	2,377.7	2,134.0	36,406.9	300
6,290.0	5,645.5	5,067.1	4,548.0	4,082.0	3,663.7	3,288.3	2,951.4	2,649.1	2,377.7	40,562.9	305
7,008.0	6,290.0	5,645.5	5,067.1	4,548.0	4,082.0	3,663.7	3,288.3	2,951.4	2,649.1	45,193.2	310
7,808.0	7,008.0	6,290.0	5,645.5	5,067.1	4,548.0	4,082.0	3,663.7	3,288.3	2,951.4	50,352.1	315
8,699.3	7,808.0	7,008.0	6,290.0	5,645.5	5,067.1	4,548.0	4,082.0	3,663.7	3,288.3	56,100.0	320
9,692.4	8,699.3	7,808.0	7,008.0	6,290.0	5,645.5	5,067.1	4,548.0	4,082.0	3,663.7	62,504.1	325
10,798.8	9,692.4	8,699.3	7,808.0	7,008.0	6,290.0	5,645.5	5,067.1	4,548.0	4,082.0	69,639.2	330
12,031.5	10,798.8	9,692.4	8,699.3	7,808.0	7,008.0	6,290.0	5,645.5	5,067.1	4,548.0	77,588.7	335
13,405.0	12,031.5	10,798.8	9,692.4	8,699.3	7,808.0	7,008.0	6,290.0	5,645.5	5,067.1	86,445.7	340
14,935.2	13,405.0	12,031.5	10,798.8	9,692.4	8,699.3	7,808.0	7,008.0	6,290.0	5,645.5	96,313.7	345
16,640.1	14,935.2	13,405.0	12,031.5	10,798.8	9,692.4	8,699.3	7,808.0	7,008.0	6,290.0	107,308.3	350
18,539.6	16,640.1	14,935.2	13,405.0	12,031.5	10,798.8	9,692.4	8,699.3	7,808.0	7,008.0	119,557.9	355
20,656.0	18,539.6	16,640.1	14,935.2	13,405.0	12,031.5	10,798.8	9,692.4	8,699.3	7,808.0	133,205.9	360
23,013.9	20,656.0	18,539.6	16,640.1	14,935.2	13,405.0	12,031.5	10,798.8	9,692.4	8,699.3	148,411.8	365
25,641.0	23,013.9	20,656.0	18,539.6	16,640.1	14,935.2	13,405.0	12,031.5	10,798.8	9,692.4	165,353.5	370
28,568.1	25,641.0	23,013.9	20,656.0	18,539.6	16,640.1	14,935.2	13,405.0	12,031.5	10,798.8	184,229.1	375
31,829.2	28,568.1	25,641.0	23,013.9	20,656.0	18,539.6	16,640.1	14,935.2	13,405.0	12,031.5	205,259.5	380
35,462.6	31,829.2	28,568.1	25,641.0	23,013.9	20,656.0	18,539.6	16,640.1	14,935.2	13,405.0	228,690.6	385
39,510.8	35,462.6	31,829.2	28,568.1	25,641.0	23,013.9	20,656.0	18,539.6	16,640.1	14,935.2	254,796.4	390
44,021.1	39,510.8	35,462.6	31,829.2	28,568.1	25,641.0	23,013.9	20,656.0	18,539.6	16,640.1	283,882.3	395
49,046.2	44,021.1	39,510.8	35,462.6	31,829.2	28,568.1	25,641.0	23,013.9	20,656.0	18,539.6	316,288.4	400
54,645.0	49,046.2	44,021.1	39,510.8	35,462.6	31,829.2	28,568.1	25,641.0	23,013.9	20,656.0	352,393.9	405
60,882.9	54,645.0	49,046.2	44,021.1	39,510.8	35,462.6	31,829.2	28,568.1	25,641.0	23,013.9	392,620.8	410
67,832.9	60,882.9	54,645.0	49,046.2	44,021.1	39,510.8	35,462.6	31,829.2	28,568.1	25,641.0	437,439.8	415
75,576.3	67,832.9	60,882.9	54,645.0	49,046.2	44,021.1	39,510.8	35,462.6	31,829.2	28,568.1	487,375.1	420
84,203.6	75,576.3	67,832.9	60,882.9	54,645.0	49,046.2	44,021.1	39,510.8	35,462.6	31,829.2	543,010.6	425
93,815.7	84,203.6	75,576.3	67,832.9	60,882.9	54,645.0	49,046.2	44,021.1	39,510.8	35,462.6	604,997.1	430

Design	Starting Population	Ending Population	Life Expectancy	Fractional	Infant Mortality	Birth Rate Required
001	70	603,550	50	Equal	No	3.584
002	66	603,550	50	Equal	No	3.598
009	70 M no W	625 M/W/C	50	Equal	No	
046	66 M no W	625 M/W/C	70	Equal	No	
113	70 M only	603 M/W/C	70	Unequal	No	
114	66 M only	603 M/W/C	70	Unequal	No	
235	70 M/W/C	625 M no W	50	Unequal	Yes	
236	66 M/W/C	625 M no W	50	Unequal	Yes	
159	70 M no W	603 M no W	50	Equal	Yes	
196	66 M no W	603 M no W	70	Equal	Yes	
203	70 M only	625 M no W	70	Equal	Yes	
204	66 M only	625 M no W	70	Equal	Yes	
097	70 M/W/C	603 M only	50	Unequal	No	
098	66 M/W/C	603 M only	50	Unequal	No	
105	70 M no W	625 M only	50	Unequal	No	
142	66 M no W	625 M only	70	Unequal	No	
065	70 M only	603 M only	70	Equal	No	
066	66 M only	603 M only	70	Equal	No	

Question: How can the reader verify the author's numerical analysis?

Answer: The author provides a sample calculation after this question and answer session.

Question: Why does the author use five year time integration?

Answer 1: The five year time integration reflects the natural process in which people have children at different stages in life. See U.S. Census Bureau, The 2007 Statistical Abstract, The National Data Book, Section 2: Vital Statistics, Table 78. Births and Birth rates by Race, Sex, and Age: 1980 to 2004, p.64

Answer 2: The five year time integration takes advantage of the power of mathematical compounding similar to the mathematical compounding in banking.

For example:

5% annual interest compounded yearly equals 1.0500

5% annual interest compounded monthly equals 1.0512

This may seem insignificant but over 430 years the difference is significant. (1.0512 divided by 1.0500) to the power of 430 equals 1.634

An increase of 63 % in 430 years.

Answer 3: The five year time integration allows for unfortunate life events like miscarriages in the pregnancy cycle without significantly effecting the numerical analysis of Israel population growth from Genesis to Exodus. Later the author will talk more about infant mortality.

Birth rate per (2) persons per (20) years — 3.584
Birth rate per (1) person per (20) years — 1.792
Birth rate per (1) person per (5) years — 0.448

Numerical analysis 1

Total starting population of 70 persons.
Total ending population of 603,550 persons
Israel stayed in Egypt 430 years.

0 to 5 Yrs	5 to 10 Yrs	10 to 15 Yrs	15 to 20 Yrs	20 to 25 Yrs	25 to 30 Yrs	30 to 35 Yrs	35 to 40 Yrs	40 to 45 Yrs	45 Yrs to 50 Yr	Total Population	Years
7.0	7.0	7.0	7.0	7.0	7.0	7.0	7.0	7.0	7.0	70.0	0
12.5	7.0	7.0	7.0	7.0	7.0	7.0	7.0	7.0	7.0	75.5	5
12.5	12.5	7.0	7.0	7.0	7.0	7.0	7.0	7.0	7.0	81.1	10
12.5	12.5	12.5	7.0	7.0	7.0	7.0	7.0	7.0	7.0	86.6	15
12.5	12.5	12.5	12.5	7.0	7.0	7.0	7.0	7.0	7.0	92.2	20
15.0	12.5	12.5	12.5	12.5	7.0	7.0	7.0	7.0	7.0	100.2	25
17.5	15.0	12.5	12.5	12.5	12.5	7.0	7.0	7.0	7.0	110.7	30
20.0	17.5	15.0	12.5	12.5	12.5	12.5	7.0	7.0	7.0	123.7	35
22.5	20.0	17.5	15.0	12.5	12.5	12.5	12.5	7.0	7.0	139.2	40
23.6	22.5	20.0	17.5	15.0	12.5	12.5	12.5	12.5	7.0	155.8	45
25.8	23.6	22.5	20.0	17.5	15.0	12.5	12.5	12.5	12.5	174.6	50
29.2	25.8	23.6	22.5	20.0	17.5	15.0	12.5	12.5	12.5	191.2	55
33.6	29.2	25.8	23.6	22.5	20.0	17.5	15.0	12.5	12.5	212.3	60
37.4	33.6	29.2	25.8	23.6	22.5	20.0	17.5	15.0	12.5	237.2	65
41.2	37.4	33.6	29.2	25.8	23.6	22.5	20.0	17.5	15.0	265.8	70
45.3	41.2	37.4	33.6	29.2	25.8	23.6	22.5	20.0	17.5	296.0	75
50.3	45.3	41.2	37.4	33.6	29.2	25.8	23.6	22.5	20.0	328.8	80
56.5	50.3	45.3	41.2	37.4	33.6	29.2	25.8	23.6	22.5	365.2	85
63.3	56.5	50.3	45.3	41.2	37.4	33.6	29.2	25.8	23.6	406.1	90
70.6	63.3	56.5	50.3	45.3	41.2	37.4	33.6	29.2	25.8	453.0	95

Sample Calculations:

Calculating New Children in Year 85 between 0 to 5 Yrs of age.

Persons in Year 80 between 15 to 20 Years	37.4
Persons in Year 80 between 20 to 25 Years	33.6
Persons in Year 80 between 25 to 30 Years	29.2
Persons in Year 80 between 30 to 35 Years	25.8
Total persons	126.0
Birth Rate per person (1) per (5) years	0.448
New children in Year 85 between 0 to 5 Yrs of age.	56.4

Machine Design Analysis 009

Starting Population

| 70 | | |

Starting Population

| | M no W | |

Ending Population

| | 625 |

Ending Population

| M/W/C | | |

Life Expectancy

| 50 | |

Fractional

| Equal | |

Infant Mortality

| No | |

For the starting population (Genesis 46: 27) assume the word "men" means males only and we need to account for females (70 M no W); therefore, the total starting population would be (70 * 2) equals 140 persons.

Assume the time the Israelites had stayed in Egypt was four hundred and thirty years (Exodus 12:40).

For the ending population (Numbers 1: 1-3, Numbers 1: 44-46, Numbers 1: 47-49, and Numbers 3:39) assume the word "men" means adult men, adult women, and children (625 M/W/C); therefore, the total ending population would be 625,550 persons.

Assume a human life expectancy of 50 years. Using numerical analysis to calculate the birth rate required for a total starting population of 140 persons to increase to a total ending population of 625,550 persons in 430 years results in a birth rate of 3.438 children.

Numerical analysis 9

Birth rate per (2) persons per (20) years	3.438
Birth rate per (1) person per (20) years	1.719
Birth rate per (1) person per (5) years	0.430

Total starting population of 140 persons.
Total ending population of 625,550 persons
Israel stayed in Egypt 430 years.

0 to 5 Yrs	5 to 10 Yrs	10 to 15 Yrs	15 to 20 Yrs	20 to 25 Yrs	25 to 30 Yrs	30 to 35 Yrs	35 to 40 Yrs	40 to 45 Yrs	45 to 50 Yrs	Total Population	Years
14.0	14.0	14.0	14.0	14.0	14.0	14.0	14.0	14.0	14.0	140.0	0
24.1	14.0	14.0	14.0	14.0	14.0	14.0	14.0	14.0	14.0	150.1	5
24.1	24.1	14.0	14.0	14.0	14.0	14.0	14.0	14.0	14.0	160.1	10
24.1	24.1	24.1	14.0	14.0	14.0	14.0	14.0	14.0	14.0	170.2	15
24.1	24.1	24.1	24.1	14.0	14.0	14.0	14.0	14.0	14.0	180.3	20
28.4	24.1	24.1	24.1	24.1	14.0	14.0	14.0	14.0	14.0	194.7	25
32.7	28.4	24.1	24.1	24.1	24.1	14.0	14.0	14.0	14.0	213.4	30
37.0	32.7	28.4	24.1	24.1	24.1	24.1	14.0	14.0	14.0	236.4	35
41.4	37.0	32.7	28.4	24.1	24.1	24.1	24.1	14.0	14.0	263.8	40
43.2	41.4	37.0	32.7	28.4	24.1	24.1	24.1	24.1	14.0	293.0	45
46.9	43.2	41.4	37.0	32.7	28.4	24.1	24.1	24.1	24.1	326.0	50
52.5	46.9	43.2	41.4	37.0	32.7	28.4	24.1	24.1	24.1	354.4	55
60.0	52.5	46.9	43.2	41.4	37.0	32.7	28.4	24.1	24.1	390.3	60
66.3	60.0	52.5	46.9	43.2	41.4	37.0	32.7	28.4	24.1	432.6	65
72.5	66.3	60.0	52.5	46.9	43.2	41.4	37.0	32.7	28.4	481.0	70
79.1	72.5	66.3	60.0	52.5	46.9	43.2	41.4	37.0	32.7	531.7	75
87.1	79.1	72.5	66.3	60.0	52.5	46.9	43.2	41.4	37.0	586.1	80
97.0	87.1	79.1	72.5	66.3	60.0	52.5	46.9	43.2	41.4	646.0	85
108.0	97.0	87.1	79.1	72.5	66.3	60.0	52.5	46.9	43.2	712.6	90
119.4	108.0	97.0	87.1	79.1	72.5	66.3	60.0	52.5	46.9	788.8	95
131.1	119.4	108.0	97.0	87.1	79.1	72.5	66.3	60.0	52.5	872.9	100
144.3	131.1	119.4	108.0	97.0	87.1	79.1	72.5	66.3	60.0	964.7	105
159.5	144.3	131.1	119.4	108.0	97.0	87.1	79.1	72.5	66.3	1,064.2	110
176.8	159.5	144.3	131.1	119.4	108.0	97.0	87.1	79.1	72.5	1,174.7	115
195.7	176.8	159.5	144.3	131.1	119.4	108.0	97.0	87.1	79.1	1,298.0	120
216.0	195.7	176.8	159.5	144.3	131.1	119.4	108.0	97.0	87.1	1,435.0	125
238.2	216.0	195.7	176.8	159.5	144.3	131.1	119.4	108.0	97.0	1,586.1	130
262.9	238.2	216.0	195.7	176.8	159.5	144.3	131.1	119.4	108.0	1,751.9	135
290.7	262.9	238.2	216.0	195.7	176.8	159.5	144.3	131.1	119.4	1,934.6	140
321.5	290.7	262.9	238.2	216.0	195.7	176.8	159.5	144.3	131.1	2,136.7	145
355.3	321.5	290.7	262.9	238.2	216.0	195.7	176.8	159.5	144.3	2,361.0	150
392.3	355.3	321.5	290.7	262.9	238.2	216.0	195.7	176.8	159.5	2,609.0	155
433.1	392.3	355.3	321.5	290.7	262.9	238.2	216.0	195.7	176.8	2,882.6	160
478.4	433.1	392.3	355.3	321.5	290.7	262.9	238.2	216.0	195.7	3,184.2	165
528.8	478.4	433.1	392.3	355.3	321.5	290.7	262.9	238.2	216.0	3,517.2	170
584.4	528.8	478.4	433.1	392.3	355.3	321.5	290.7	262.9	238.2	3,885.5	175
645.6	584.4	528.8	478.4	433.1	392.3	355.3	321.5	290.7	262.9	4,292.9	180
713.0	645.6	584.4	528.8	478.4	433.1	392.3	355.3	321.5	290.7	4,743.0	185
787.5	713.0	645.6	584.4	528.8	478.4	433.1	392.3	355.3	321.5	5,239.9	190
870.1	787.5	713.0	645.6	584.4	528.8	478.4	433.1	392.3	355.3	5,788.5	195
961.4	870.1	787.5	713.0	645.6	584.4	528.8	478.4	433.1	392.3	6,394.5	200
1,062.2	961.4	870.1	787.5	713.0	645.6	584.4	528.8	478.4	433.1	7,064.5	205

0 to 5 Yrs	5 to 10 Yrs	10 to 15 Yrs	15 to 20 Yrs	20 to 25 Yrs	25 to 30 Yrs	30 to 35 Yrs	35 to 40 Yrs	40 to 45 Yrs	45 to 50 Yrs	Total Population	Years
1,173.4	1,062.2	961.4	870.1	787.5	713.0	645.6	584.4	528.8	478.4	7,804.8	210
1,296.2	1,173.4	1,062.2	961.4	870.1	787.5	713.0	645.6	584.4	528.8	8,622.6	215
1,431.9	1,296.2	1,173.4	1,062.2	961.4	870.1	787.5	713.0	645.6	584.4	9,525.8	220
1,582.0	1,431.9	1,296.2	1,173.4	1,062.2	961.4	870.1	787.5	713.0	645.6	10,523.5	225
1,747.9	1,582.0	1,431.9	1,296.2	1,173.4	1,062.2	961.4	870.1	787.5	713.0	11,625.8	230
1,931.0	1,747.9	1,582.0	1,431.9	1,296.2	1,173.4	1,062.2	961.4	870.1	787.5	12,843.7	235
2,133.2	1,931.0	1,747.9	1,582.0	1,431.9	1,296.2	1,173.4	1,062.2	961.4	870.1	14,189.4	240
2,356.6	2,133.2	1,931.0	1,747.9	1,582.0	1,431.9	1,296.2	1,173.4	1,062.2	961.4	15,675.9	245
2,603.4	2,356.6	2,133.2	1,931.0	1,747.9	1,582.0	1,431.9	1,296.2	1,173.4	1,062.2	17,317.9	250
2,876.2	2,603.4	2,356.6	2,133.2	1,931.0	1,747.9	1,582.0	1,431.9	1,296.2	1,173.4	19,131.9	255
3,177.6	2,876.2	2,603.4	2,356.6	2,133.2	1,931.0	1,747.9	1,582.0	1,431.9	1,296.2	21,136.1	260
3,510.5	3,177.6	2,876.2	2,603.4	2,356.6	2,133.2	1,931.0	1,747.9	1,582.0	1,431.9	23,350.4	265
3,878.2	3,510.5	3,177.6	2,876.2	2,603.4	2,356.6	2,133.2	1,931.0	1,747.9	1,582.0	25,796.6	270
4,284.4	3,878.2	3,510.5	3,177.6	2,876.2	2,603.4	2,356.6	2,133.2	1,931.0	1,747.9	28,498.9	275
4,733.2	4,284.4	3,878.2	3,510.5	3,177.6	2,876.2	2,603.4	2,356.6	2,133.2	1,931.0	31,484.3	280
5,229.1	4,733.2	4,284.4	3,878.2	3,510.5	3,177.6	2,876.2	2,603.4	2,356.6	2,133.2	34,782.4	285
5,776.9	5,229.1	4,733.2	4,284.4	3,878.2	3,510.5	3,177.6	2,876.2	2,603.4	2,356.6	38,426.1	290
6,382.1	5,776.9	5,229.1	4,733.2	4,284.4	3,878.2	3,510.5	3,177.6	2,876.2	2,603.4	42,451.6	295
7,050.6	6,382.1	5,776.9	5,229.1	4,733.2	4,284.4	3,878.2	3,510.5	3,177.6	2,876.2	46,898.7	300
7,789.2	7,050.6	6,382.1	5,776.9	5,229.1	4,733.2	4,284.4	3,878.2	3,510.5	3,177.6	51,811.6	305
8,605.1	7,789.2	7,050.6	6,382.1	5,776.9	5,229.1	4,733.2	4,284.4	3,878.2	3,510.5	57,239.1	310
9,506.6	8,605.1	7,789.2	7,050.6	6,382.1	5,776.9	5,229.1	4,733.2	4,284.4	3,878.2	63,235.3	315
10,502.5	9,506.6	8,605.1	7,789.2	7,050.6	6,382.1	5,776.9	5,229.1	4,733.2	4,284.4	69,859.6	320
11,602.7	10,502.5	9,506.6	8,605.1	7,789.2	7,050.6	6,382.1	5,776.9	5,229.1	4,733.2	77,177.9	325
12,818.1	11,602.7	10,502.5	9,506.6	8,605.1	7,789.2	7,050.6	6,382.1	5,776.9	5,229.1	85,262.8	330
14,160.9	12,818.1	11,602.7	10,502.5	9,506.6	8,605.1	7,789.2	7,050.6	6,382.1	5,776.9	94,194.6	335
15,644.4	14,160.9	12,818.1	11,602.7	10,502.5	9,506.6	8,605.1	7,789.2	7,050.6	6,382.1	104,062.1	340
17,283.2	15,644.4	14,160.9	12,818.1	11,602.7	10,502.5	9,506.6	8,605.1	7,789.2	7,050.6	114,963.3	345
19,093.8	17,283.2	15,644.4	14,160.9	12,818.1	11,602.7	10,502.5	9,506.6	8,605.1	7,789.2	127,006.5	350
21,093.9	19,093.8	17,283.2	15,644.4	14,160.9	12,818.1	11,602.7	10,502.5	9,506.6	8,605.1	140,311.2	355
23,303.7	21,093.9	19,093.8	17,283.2	15,644.4	14,160.9	12,818.1	11,602.7	10,502.5	9,506.6	155,009.7	360
25,744.9	23,303.7	21,093.9	19,093.8	17,283.2	15,644.4	14,160.9	12,818.1	11,602.7	10,502.5	171,248.0	365
28,441.8	25,744.9	23,303.7	21,093.9	19,093.8	17,283.2	15,644.4	14,160.9	12,818.1	11,602.7	189,187.3	370
31,421.3	28,441.8	25,744.9	23,303.7	21,093.9	19,093.8	17,283.2	15,644.4	14,160.9	12,818.1	209,005.9	375
34,712.9	31,421.3	28,441.8	25,744.9	23,303.7	21,093.9	19,093.8	17,283.2	15,644.4	14,160.9	230,900.7	380
38,349.3	34,712.9	31,421.3	28,441.8	25,744.9	23,303.7	21,093.8	19,093.8	17,283.2	15,644.4	255,089.0	385
42,366.6	38,349.3	34,712.9	31,421.3	28,441.8	25,744.9	23,303.7	21,093.9	19,093.8	17,283.2	281,811.3	390
46,804.8	42,366.6	38,349.3	34,712.9	31,421.3	28,441.8	25,744.9	23,303.7	21,093.9	19,093.8	311,332.8	395
51,707.9	46,804.8	42,366.6	38,349.3	34,712.9	31,421.3	28,441.8	25,744.9	23,303.7	21,093.9	343,946.9	400
57,124.6	51,707.9	46,804.8	42,366.6	38,349.3	34,712.9	31,421.3	28,441.8	25,744.9	23,303.7	379,977.6	405
63,108.8	57,124.6	51,707.9	46,804.8	42,366.6	38,349.3	34,712.9	31,421.3	28,441.8	25,744.9	419,782.8	410
69,719.8	63,108.8	57,124.6	51,707.9	46,804.8	42,366.6	38,349.3	34,712.9	31,421.3	28,441.8	463,757.8	415
77,023.5	69,719.8	63,108.8	57,124.6	51,707.9	46,804.8	42,366.6	38,349.3	34,712.9	31,421.3	512,339.4	420
85,092.2	77,023.5	69,719.8	63,108.8	57,124.6	51,707.9	46,804.8	42,366.6	38,349.3	34,712.9	566,010.3	425
94,006.1	85,092.2	77,023.5	69,719.8	63,108.8	57,124.6	51,707.9	46,804.8	42,366.6	38,349.3	625,303.5	430

Design	Starting Population	Ending Population	Life Expectancy	Fractional	Infant Mortality	Birth Rate Required
001	70	603,550	50	Equal	No	3.584
002	66	603,550	50	Equal	No	3.598
009	140	625,550	50	Equal	No	3.438
046	66 M no W	625 M/W/C	70	Equal	No	
113	70 M only	603 M/W/C	70	Unequal	No	
114	66 M only	603 M/W/C	70	Unequal	No	
235	70 M/W/C	625 M no W	50	Unequal	Yes	
236	66 M/W/C	625 M no W	50	Unequal	Yes	
159	70 M no W	603 M no W	50	Equal	Yes	
196	66 M no W	603 M no W	70	Equal	Yes	
203	70 M only	625 M no W	50	Equal	Yes	
204	66 M only	625 M no W	70	Equal	Yes	
097	70 M/W/C	603 M only	50	Unequal	No	
098	66 M/W/C	603 M only	50	Unequal	No	
105	70 M no W	625 M only	50	Unequal	No	
142	66 M no W	625 M only	70	Unequal	No	
065	70 M only	603 M only	70	Equal	No	
066	66 M only	603 M only	70	Equal	No	

Question: What does the author mean by equal fractional?

Answer: For Israel population growth from Genesis to Exodus equal fractional assumes women bearing children occurs equally over time.

25 % of the children are born to women between 15 to 20 years of age

25 % of the children are born to women between 20 to 25 years of age

25 % of the children are born to women between 25 to 30 years of age

25 % of the children are born to women between 30 to 35 years of age

Later the author will also use unequal fractional (adjusted for U.S. Census Bureau Data). This book uses equal fractional and unequal fractional as the two possibilities when analyzing Israel population growth from Genesis to Exodus.

Machine Design Analysis 046

Starting Population

	66

Starting Population

	M no W	

Ending Population

	625

Ending Population

M/W/C		

Life Expectancy

	70

Fractional

Equal	

Infant Mortality

No	

For the starting population (Genesis 46: 26-27) assume the word "men" means males only and we need to account for females (66 M no W); therefore, the total starting population would be (66 * 2) equals 132 persons.

Assume the time the Israelites had stayed in Egypt was four hundred and thirty years (Exodus 12:40).

For the ending population (Numbers 1: 1-3, Numbers 1: 44-46, Numbers 1: 47-49, and Numbers 3:39) assume the word "men" means adult men, adult women, and children (625 M/W/C); therefore, the total ending population would be 625,550 persons.

Assume a human life expectancy of 70 years (Psalms 90: 8-10. Using numerical analysis to calculate the birth rate required for a total starting population of 132 persons to increase to a total ending population of 625,550 persons in 430 years results in a birth rate of 3.487 children.

Stanley R. Stasko

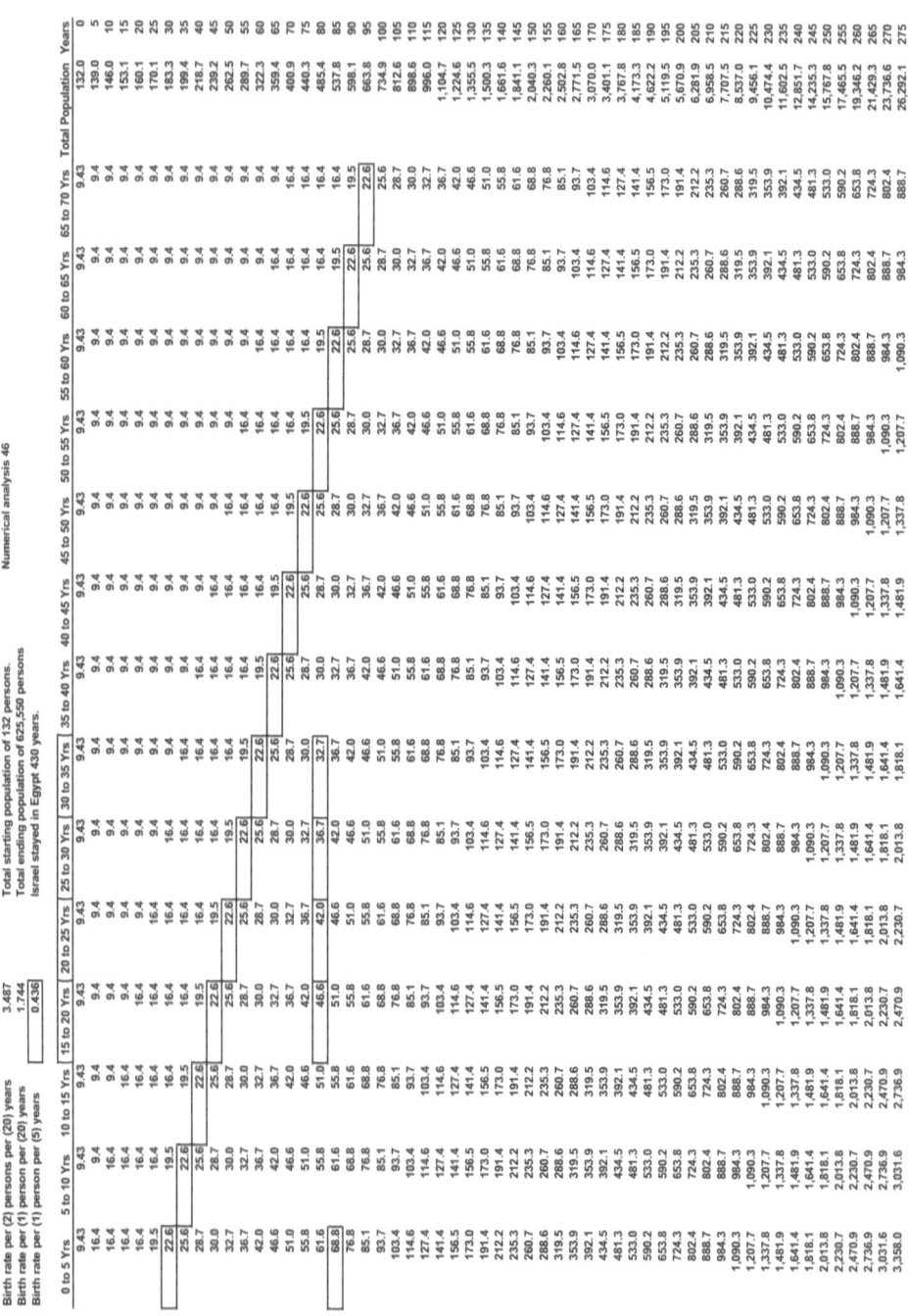

0 to 5 Yrs	5 to 10 Yrs	10 to 15 Yrs	15 to 20 Yrs	20 to 25 Yrs	25 to 30 Yrs	30 to 35 Yrs	35 to 40 Yrs	40 to 45 Yrs	45 to 50 Yrs	50 to 55 Yrs	55 to 60 Yrs	60 to 65 Yrs	65 to 70 Yrs	Total Population	Years
3,719.5	3,358.0	3,031.6	2,736.9	2,470.9	2,230.7	2,013.8	1,818.1	1,641.4	1,481.9	1,337.8	1,207.7	1,090.3	984.3	29,122.9	280
4,120.0	3,719.5	3,358.0	3,031.6	2,736.9	2,470.9	2,230.7	2,013.8	1,818.1	1,641.4	1,461.9	1,337.8	1,207.7	1,090.3	32,258.6	285
4,563.7	4,120.0	3,719.5	3,358.0	3,031.6	2,736.9	2,470.9	2,230.7	2,013.8	1,818.1	1,641.4	1,481.9	1,337.8	1,207.7	35,732.0	290
5,055.0	4,563.7	4,120.0	3,719.5	3,358.0	3,031.6	2,736.9	2,470.9	2,230.7	2,013.8	1,818.1	1,641.4	1,461.9	1,337.8	39,579.3	295
5,599.3	5,055.0	4,563.7	4,120.0	3,719.5	3,358.0	3,031.6	2,736.9	2,470.9	2,230.7	2,013.8	1,818.1	1,641.4	1,461.9	43,840.7	300
6,202.1	5,599.3	5,055.0	4,563.7	4,120.0	3,719.5	3,358.0	3,031.6	2,736.9	2,470.9	2,230.7	2,013.8	1,818.1	1,641.4	48,560.9	305
6,869.9	6,202.1	5,599.3	5,055.0	4,563.7	4,120.0	3,719.5	3,358.0	3,031.6	2,736.9	2,470.9	2,230.7	2,013.8	1,818.1	53,789.4	310
7,609.6	6,869.9	6,202.1	5,599.3	5,055.0	4,563.7	4,120.0	3,719.5	3,358.0	3,031.6	2,736.9	2,470.9	2,230.7	2,013.8	59,581.0	315
8,428.9	7,609.6	6,869.9	6,202.1	5,599.3	5,055.0	4,563.7	4,120.0	3,719.5	3,358.0	3,031.6	2,736.9	2,470.9	2,230.7	65,996.1	320
9,336.5	8,428.9	7,609.6	6,869.9	6,202.1	5,599.3	5,055.0	4,563.7	4,120.0	3,719.5	3,358.0	3,031.6	2,736.9	2,470.9	73,101.9	325
10,341.7	9,336.5	8,428.9	7,609.6	6,869.9	6,202.1	5,599.3	5,055.0	4,563.7	4,120.0	3,719.5	3,358.0	3,031.6	2,736.9	80,972.7	330
11,455.2	10,341.7	9,336.5	8,428.9	7,609.6	6,869.9	6,202.1	5,599.3	5,055.0	4,563.7	4,120.0	3,719.5	3,358.0	3,031.6	89,690.9	335
12,688.6	11,455.2	10,341.7	9,336.5	8,428.9	7,609.6	6,869.9	6,202.1	5,599.3	5,055.0	4,563.7	4,120.0	3,719.5	3,358.0	99,347.9	340
14,054.7	12,688.6	11,455.2	10,341.7	9,336.5	8,428.9	7,609.6	6,869.9	6,202.1	5,599.3	5,055.0	4,563.7	4,120.0	3,719.5	110,044.7	345
15,568.0	14,054.7	12,688.6	11,455.2	10,341.7	9,336.5	8,428.9	7,609.6	6,869.9	6,202.1	5,599.3	5,055.0	4,563.7	4,120.0	121,893.2	350
17,244.2	15,568.0	14,054.7	12,688.6	11,455.2	10,341.7	9,336.5	8,428.9	7,609.6	6,869.9	6,202.1	5,599.3	5,055.0	4,563.7	135,017.3	355
19,100.9	17,244.2	15,568.0	14,054.7	12,688.6	11,455.2	10,341.7	9,336.5	8,428.9	7,609.6	6,869.9	6,202.1	5,599.3	5,055.0	149,544.5	360
21,157.4	19,100.9	17,244.2	15,568.0	14,054.7	12,688.6	11,455.2	10,341.7	9,336.5	8,428.9	7,609.6	6,869.9	6,202.1	5,599.3	165,657.0	365
23,435.5	21,157.4	19,100.9	17,244.2	15,568.0	14,054.7	12,688.6	11,455.2	10,341.7	9,336.5	8,428.9	7,609.6	6,869.9	6,202.1	183,493.2	370
25,958.8	23,435.5	21,157.4	19,100.9	17,244.2	15,568.0	14,054.7	12,688.6	11,455.2	10,341.7	9,336.5	8,428.9	7,609.6	6,869.9	203,249.8	375
28,753.7	25,958.8	23,435.5	21,157.4	19,100.9	17,244.2	15,568.0	14,054.7	12,688.6	11,455.2	10,341.7	9,336.5	8,428.9	7,609.6	225,133.7	380
31,849.6	28,753.7	25,958.8	23,435.5	21,157.4	19,100.9	17,244.2	15,568.0	14,054.7	12,688.6	11,455.2	10,341.7	9,336.5	8,428.9	249,373.7	385
35,278.8	31,849.6	28,753.7	25,958.8	23,435.5	21,157.4	19,100.9	17,244.2	15,568.0	14,054.7	12,688.6	11,455.2	10,341.7	9,336.5	276,223.6	390
39,077.3	35,278.8	31,849.6	28,753.7	25,958.8	23,435.5	21,157.4	19,100.9	17,244.2	15,568.0	14,054.7	12,688.6	11,455.2	10,341.7	305,964.4	395
43,284.7	39,077.3	35,278.8	31,849.6	28,753.7	25,958.8	23,435.5	21,157.4	19,100.9	17,244.2	15,568.0	14,054.7	12,688.6	11,455.2	338,907.5	400
47,945.2	43,284.7	39,077.3	35,278.8	31,849.6	28,753.7	25,958.8	23,435.5	21,157.4	19,100.9	17,244.2	15,568.0	14,054.7	12,688.6	375,397.5	405
53,107.4	47,945.2	43,284.7	39,077.3	35,278.8	31,849.6	28,753.7	25,958.8	23,435.5	21,157.4	19,100.9	17,244.2	15,568.0	14,054.7	415,816.4	410
58,825.5	53,107.4	47,945.2	43,284.7	39,077.3	35,278.8	31,849.6	28,753.7	25,958.8	23,435.5	21,157.4	19,100.9	17,244.2	15,568.0	460,587.1	415
65,159.2	58,825.5	53,107.4	47,945.2	43,284.7	39,077.3	35,278.8	31,849.6	28,753.7	25,958.8	23,435.5	21,157.4	19,100.9	17,244.2	510,178.2	420
72,174.8	65,159.2	58,825.5	53,107.4	47,945.2	43,284.7	39,077.3	35,278.8	31,849.6	28,753.7	25,958.8	23,435.5	21,157.4	19,100.9	565,108.9	425
79,945.9	72,174.8	65,159.2	58,825.5	53,107.4	47,945.2	43,284.7	39,077.3	35,278.8	31,849.6	28,753.7	25,958.8	23,435.5	21,157.4	625,953.9	430

Design	Starting Population	Ending Population	Life Expectancy	Fractional	Infant Mortality	Birth Rate Required
001	70	603,550	50	Equal	No	3.584
002	66	603,550	50	Equal	No	3.598
009	140	625,550	50	Equal	No	3.438
046	132	625,550	70	Equal	No	3.487
113	70 M only	603 M/W/C	70	Unequal	No	
114	66 M only	603 M/W/C	70	Unequal	No	
235	70 M/W/C	625 M no W	50	Unequal	Yes	
236	66 M/W/C	625 M no W	50	Unequal	Yes	
159	70 M no W	603 M no W	50	Equal	Yes	
196	66 M no W	603 M no W	70	Equal	Yes	
203	70 M only	625 M no W	70	Equal	Yes	
204	66 M only	625 M no W	70	Equal	Yes	
097	70 M/W/C	603 M only	50	Unequal	No	
098	66 M/W/C	603 M only	50	Unequal	No	
105	70 M no W	625 M only	50	Unequal	No	
142	66 M no W	625 M only	70	Unequal	No	
065	70 M only	603 M only	70	Equal	No	
066	66 M only	603 M only	70	Equal	No	

Question: What does the author mean by unequal fractional?

Answer: For Israel population growth from Genesis to Exodus unequal fractional assumes women bearing children occurs similar to U.S. Census Bureau Data over time.

12.5 % of the children are born to women under 20 years of age

27.0 % of the children are born to women between 20 to 24 years of age

28.6 % of the children are born to women between 25 to 29 years of age

22.2 % of the children are born to women between 30 to 34 years of age

9.6 % of the children are born to women between 35 to 39 years of age

U.S. Census Bureau, The 2007 Statistical Abstract

The National Data Book, Section 2: Vital Statistics

Table 78. Births and Birth rates by Race, Sex, and Age: 1980 to 2004, p.64

Live Births | X 1000

Age of Mother	1980	1985	1990	1995	1998	1999	2000	2001	2002	2003	Totals
Under 20 Years Old	562	478	533	512	494	485	478	454	433	421	4,850
20 to 24 Years Old	1,226	1,141	1,094	966	965	982	1,018	1,022	1,022	1,032	10,468
25 to 29 Years Old	1,108	1,201	1,277	1,064	1,083	1,078	1,088	1,058	1,060	1,086	11,103
30 to 34 Years Old	550	696	886	905	889	892	929	943	951	976	8,617
35 to 39 Years Old	141	214	318	384	425	434	452	452	454	468	3,742
											38,780

Summary Calculations

Live Births | X 1000

Age of Mother	Totals	Fraction	Percentage
Under 20 Years Old	4,850	0.125	12.5%
20 to 24 Years Old	10,468	0.270	27.0%
25 to 29 Years Old	11,103	0.286	28.6%
30 to 34 Years Old	8,617	0.222	22.2%
35 to 39 Years Old	3,742	0.096	9.6%
	38,780		

Question: How can the reader verify the author's calculations for unequal fractional?

Answer: The author provides a sample calculation for unequal fractional after machine design analysis 113.

Machine Design Analysis 113

Starting Population

| 70 | | |

Starting Population

| | | M only |

Ending Population

| 603 | | |

Ending Population

| M/W/C | | |

Life Expectancy

| | 70 |

Fractional

| | Unequal |

Infant Mortality

| No | |

For the starting population (Genesis 46: 27) assume the word "men" means adult males only and we need to account for adult females, male children, and female children (70 M only); therefore, the total starting population would be (70 * 2) + (70 * 3.366) equals 376 persons.

Assume the time the Israelites had stayed in Egypt was four hundred and thirty years (Exodus 12:40).

For the ending population (Numbers 1: 1-3, Numbers 1: 44-46) assume the word "men" means adult men, adult women, and children (603 M/W/C); therefore, the total ending population would be 603,550 persons.

Assume a human life expectancy of 70 years (Psalms 90: 8-10. Using numerical analysis to calculate the birth rate required for a total starting population of 376 persons to increase to a total ending population of 603,550 persons in 430 years results in a birth rate of 3.366 children.

Numerical analysis 113

Birth rate per (2) persons per (25) years   3.366
Birth rate per (1) person per (25) years   1.683
Birth rate per (1) person per (5) years   0.337

Total starting population of 376 persons.
Total ending population 603,550 persons.
Israel stayed in Egypt 430 years.

Age-group fertility fractions: 15 to 20 Yrs = 0.125; 20 to 25 Yrs = 0.270; 25 to 30 Yrs = 0.286; 30 to 35 Yrs = 0.222; 35 to 40 Yrs = 0.096

0 to 5 Yrs	5 to 10 Yrs	10 to 15 Yrs	15 to 20 Yrs	20 to 25 Yrs	25 to 30 Yrs	30 to 35 Yrs	35 to 40 Yrs	40 to 45 Yrs	45 to 50 Yrs	50 to 55 Yrs	55 to 60 Yrs	60 to 65 Yrs	65 to 70 Yrs	Total Population	Years
26.86	26.9	26.9	26.9	26.9	26.9	26.9	26.9	26.9	26.9	26.9	26.9	26.9	26.9	376.0	0
45.2	26.9	26.9	26.9	26.9	26.9	26.9	26.9	26.9	26.9	26.9	26.9	26.9	26.9	394.3	5
45.2	45.2	26.9	26.9	26.9	26.9	26.9	26.9	26.9	26.9	26.9	26.9	26.9	26.9	412.6	10
45.2	45.2	45.2	26.9	26.9	26.9	26.9	26.9	26.9	26.9	26.9	26.9	26.9	26.9	430.9	15
45.2	45.2	45.2	45.2	26.9	26.9	26.9	26.9	26.9	26.9	26.9	26.9	26.9	26.9	449.2	20
49.0	45.2	45.2	45.2	45.2	26.9	26.9	26.9	26.9	26.9	26.9	26.9	26.9	26.9	471.4	25
57.3	49.0	45.2	45.2	45.2	45.2	26.9	26.9	26.9	26.9	26.9	26.9	26.9	26.9	501.9	30
66.1	57.3	49.0	45.2	45.2	45.2	45.2	26.9	26.9	26.9	26.9	26.9	26.9	26.9	541.1	35
73.0	66.1	57.3	49.0	45.2	45.2	45.2	45.2	26.9	26.9	26.9	26.9	26.9	26.9	587.2	40
76.7	73.0	66.1	57.3	49.0	45.2	45.2	45.2	45.2	26.9	26.9	26.9	26.9	26.9	637.1	45
80.2	76.7	73.0	66.1	57.3	49.0	45.2	45.2	45.2	45.2	26.9	26.9	26.9	26.9	690.5	50
87.7	80.2	76.7	73.0	66.1	57.3	49.0	45.2	45.2	45.2	45.2	26.9	26.9	26.9	751.4	55
98.6	87.7	80.2	76.7	73.0	66.1	57.3	49.0	45.2	45.2	45.2	45.2	26.9	26.9	823.1	60
110.5	98.6	87.7	80.2	76.7	73.0	66.1	57.3	49.0	45.2	45.2	45.2	45.2	26.9	906.7	65
120.8	110.5	98.6	87.7	80.2	76.7	73.0	66.1	57.3	49.0	45.2	45.2	45.2	45.2	1,000.7	70
129.8	120.8	110.5	98.6	87.7	80.2	76.7	73.0	66.1	57.3	49.0	45.2	45.2	45.2	1,085.3	75
139.7	129.8	120.8	110.5	98.6	87.7	80.2	76.7	73.0	66.1	57.3	49.0	45.2	45.2	1,179.9	80
152.6	139.7	129.8	120.8	110.5	98.6	87.7	80.2	76.7	73.0	66.1	57.3	49.0	45.2	1,287.4	85
168.8	152.6	139.7	129.8	120.8	110.5	98.6	87.7	80.2	76.7	73.0	66.1	57.3	49.0	1,411.0	90
186.4	168.8	152.6	139.7	129.8	120.8	110.5	98.6	87.7	80.2	76.7	73.0	66.1	57.3	1,548.4	95
203.7	186.4	168.8	152.6	139.7	129.8	120.8	110.5	98.6	87.7	80.2	76.7	73.0	66.1	1,694.8	100
221.1	203.7	186.4	168.8	152.6	139.7	129.8	120.8	110.5	98.6	87.7	80.2	76.7	73.0	1,849.8	105
240.1	221.1	203.7	186.4	168.8	152.6	139.7	129.8	120.8	110.5	98.6	87.7	80.2	76.7	2,017.0	110
262.6	240.1	221.1	203.7	186.4	168.8	152.6	139.7	129.8	120.8	110.5	98.6	87.7	80.2	2,202.8	115
288.4	262.6	240.1	221.1	203.7	186.4	168.8	152.6	139.7	129.8	120.8	110.5	98.6	87.7	2,411.0	120
316.6	288.4	262.6	240.1	221.1	203.7	186.4	168.8	152.6	139.7	129.8	120.8	110.5	98.6	2,639.8	125
346.0	316.6	288.4	262.6	240.1	221.1	203.7	186.4	168.8	152.6	139.7	129.8	120.8	110.5	2,887.2	130
377.0	346.0	316.6	288.4	262.6	240.1	221.1	203.7	186.4	168.8	152.6	139.7	129.8	120.8	3,153.7	135
411.1	377.0	346.0	316.6	288.4	262.6	240.1	221.1	203.7	186.4	168.8	152.6	139.7	129.8	3,444.0	140
449.5	411.1	377.0	346.0	316.6	288.4	262.6	240.1	221.1	203.7	186.4	168.8	152.6	139.7	3,763.7	145
492.4	449.5	411.1	377.0	346.0	316.6	288.4	262.6	240.1	221.1	203.7	186.4	168.8	152.6	4,116.3	150
539.1	492.4	449.5	411.1	377.0	346.0	316.6	288.4	262.6	240.1	221.1	203.7	186.4	168.8	4,502.8	155
589.2	539.1	492.4	449.5	411.1	377.0	346.0	316.6	288.4	262.6	240.1	221.1	203.7	186.4	4,923.2	160
643.2	589.2	539.1	492.4	449.5	411.1	377.0	346.0	316.6	288.4	262.6	240.1	221.1	203.7	5,380.0	165
702.5	643.2	589.2	539.1	492.4	449.5	411.1	377.0	346.0	316.6	288.4	262.6	240.1	221.1	5,878.7	170
768.0	702.5	643.2	589.2	539.1	492.4	449.5	411.1	377.0	346.0	316.6	288.4	262.6	240.1	6,425.7	175
840.3	768.0	702.5	643.2	589.2	539.1	492.4	449.5	411.1	377.0	346.0	316.6	288.4	262.6	7,025.8	180
919.1	840.3	768.0	702.5	643.2	589.2	539.1	492.4	449.5	411.1	377.0	346.0	316.6	288.4	7,682.4	185
1,004.7	919.1	840.3	768.0	702.5	643.2	589.2	539.1	492.4	449.5	411.1	377.0	346.0	316.6	8,398.6	190
1,097.6	1,004.7	919.1	840.3	768.0	702.5	643.2	589.2	539.1	492.4	449.5	411.1	377.0	346.0	9,179.7	195
1,199.4	1,097.6	1,004.7	919.1	840.3	768.0	702.5	643.2	589.2	539.1	492.4	449.5	411.1	377.0	10,033.1	200
1,311.3	1,199.4	1,097.6	1,004.7	919.1	840.3	768.0	702.5	643.2	589.2	539.1	492.4	449.5	411.1	10,967.4	205
1,433.9	1,311.3	1,199.4	1,097.6	1,004.7	919.1	840.3	768.0	702.5	643.2	589.2	539.1	492.4	449.5	11,990.2	210
1,567.9	1,433.9	1,311.3	1,199.4	1,097.6	1,004.7	919.1	840.3	768.0	702.5	643.2	589.2	539.1	492.4	13,106.6	215
1,713.9	1,567.9	1,433.9	1,311.3	1,199.4	1,097.6	1,004.7	919.1	840.3	768.0	702.5	643.2	589.2	539.1	14,330.1	220
1,873.1	1,713.9	1,567.9	1,433.9	1,311.3	1,199.4	1,097.6	1,004.7	919.1	840.3	768.0	702.5	643.2	589.2	15,664.1	225
2,047.3	1,873.1	1,713.9	1,567.9	1,433.9	1,311.3	1,199.4	1,097.6	1,004.7	919.1	840.3	768.0	702.5	643.2	17,122.1	230
2,238.1	2,047.3	1,873.1	1,713.9	1,567.9	1,433.9	1,311.3	1,199.4	1,097.6	1,004.7	919.1	840.3	768.0	702.5	18,717.0	235
2,446.9	2,238.1	2,047.3	1,873.1	1,713.9	1,567.9	1,433.9	1,311.3	1,199.4	1,097.6	1,004.7	919.1	840.3	768.0	20,461.4	240
2,675.1	2,446.9	2,238.1	2,047.3	1,873.1	1,713.9	1,567.9	1,433.9	1,311.3	1,199.4	1,097.6	1,004.7	919.1	840.3	22,368.5	245
2,924.3	2,675.1	2,446.9	2,238.1	2,047.3	1,873.1	1,713.9	1,567.9	1,433.9	1,311.3	1,199.4	1,097.6	1,004.7	919.1	24,452.5	250
3,196.4	2,924.3	2,675.1	2,446.9	2,238.1	2,047.3	1,873.1	1,713.9	1,567.9	1,433.9	1,311.3	1,199.4	1,097.6	1,004.7	26,729.8	255
3,493.9	3,196.4	2,924.3	2,675.1	2,446.9	2,238.1	2,047.3	1,873.1	1,713.9	1,567.9	1,433.9	1,311.3	1,199.4	1,097.6	29,219.1	260
3,819.5	3,493.9	3,196.4	2,924.3	2,675.1	2,446.9	2,238.1	2,047.3	1,873.1	1,713.9	1,567.9	1,433.9	1,311.3	1,199.4	31,940.9	265
4,175.6	3,819.5	3,493.9	3,196.4	2,924.3	2,675.1	2,446.9	2,238.1	2,047.3	1,873.1	1,713.9	1,567.9	1,433.9	1,311.3	34,917.1	270

Israel Population Growth

4,564.7	4,175.6	3,819.5	3,493.9	3,196.4	2,924.3	2,675.1	2,446.9	2,238.1	2,047.3	1,873.1	1,713.9	1,567.9	1,433.9	38,170.6	275
4,989.9	4,564.7	4,175.6	3,819.5	3,493.9	3,196.4	2,924.3	2,675.1	2,446.9	2,238.1	2,047.3	1,873.1	1,713.9	1,567.9	41,726.5	280
5,454.5	4,989.9	4,564.7	4,175.6	3,819.5	3,493.9	3,196.4	2,924.3	2,675.1	2,446.9	2,238.1	2,047.3	1,873.1	1,713.9	45,613.2	285
5,962.5	5,454.5	4,989.9	4,564.7	4,175.6	3,819.5	3,493.9	3,196.4	2,924.3	2,675.1	2,446.9	2,238.1	2,047.3	1,873.1	49,861.9	290
6,518.1	5,962.5	5,454.5	4,989.9	4,564.7	4,175.6	3,819.5	3,493.9	3,196.4	2,924.3	2,675.1	2,446.9	2,238.1	2,047.3	54,506.9	295
7,125.5	6,518.1	5,962.5	5,454.5	4,989.9	4,564.7	4,175.6	3,819.5	3,493.9	3,196.4	2,924.3	2,675.1	2,446.9	2,238.1	59,585.0	300
7,789.4	7,125.5	6,518.1	5,962.5	5,454.5	4,989.9	4,564.7	4,175.6	3,819.5	3,493.9	3,196.4	2,924.3	2,675.1	2,446.9	65,136.3	305
8,514.9	7,789.4	7,125.5	6,518.1	5,962.5	5,454.5	4,989.9	4,564.7	4,175.6	3,819.5	3,493.9	3,196.4	2,924.3	2,675.1	71,204.4	310
9,308.0	8,514.9	7,789.4	7,125.5	6,518.1	5,962.5	5,454.5	4,989.9	4,564.7	4,175.6	3,819.5	3,493.9	3,196.4	2,924.3	77,837.3	315
10,175.1	9,308.0	8,514.9	7,789.4	7,125.5	6,518.1	5,962.5	5,454.5	4,989.9	4,564.7	4,175.6	3,819.5	3,493.9	3,196.4	85,088.0	320
11,123.0	10,175.1	9,308.0	8,514.9	7,789.4	7,125.5	6,518.1	5,962.5	5,454.5	4,989.9	4,564.7	4,175.6	3,819.5	3,493.9	93,014.7	325
12,159.3	11,123.0	10,175.1	9,308.0	8,514.9	7,789.4	7,125.5	6,518.1	5,962.5	5,454.5	4,989.9	4,564.7	4,175.6	3,819.5	101,680.0	330
13,292.1	12,159.3	11,123.0	10,175.1	9,308.0	8,514.9	7,789.4	7,125.5	6,518.1	5,962.5	5,454.5	4,989.9	4,564.7	4,175.6	111,152.7	335
14,530.4	13,292.1	12,159.3	11,123.0	10,175.1	9,308.0	8,514.9	7,789.4	7,125.5	6,518.1	5,962.5	5,454.5	4,989.9	4,564.7	121,507.5	340
15,883.9	14,530.4	13,292.1	12,159.3	11,123.0	10,175.1	9,308.0	8,514.9	7,789.4	7,125.5	6,518.1	5,962.5	5,454.5	4,989.9	132,826.6	345
17,363.5	15,883.9	14,530.4	13,292.1	12,159.3	11,123.0	10,175.1	9,308.0	8,514.9	7,789.4	7,125.5	6,518.1	5,962.5	5,454.5	145,200.2	350
18,981.1	17,363.5	15,883.9	14,530.4	13,292.1	12,159.3	11,123.0	10,175.1	9,308.0	8,514.9	7,789.4	7,125.5	6,518.1	5,962.5	158,726.8	355
20,749.4	18,981.1	17,363.5	15,883.9	14,530.4	13,292.1	12,159.3	11,123.0	10,175.1	9,308.0	8,514.9	7,789.4	7,125.5	6,518.1	173,513.7	360
22,682.5	20,749.4	18,981.1	17,363.5	15,883.9	14,530.4	13,292.1	12,159.3	11,123.0	10,175.1	9,308.0	8,514.9	7,789.4	7,125.5	189,678.1	365
24,795.5	22,682.5	20,749.4	18,981.1	17,363.5	15,883.9	14,530.4	13,292.1	12,159.3	11,123.0	10,175.1	9,308.0	8,514.9	7,789.4	207,348.1	370
27,105.3	24,795.5	22,682.5	20,749.4	18,981.1	17,363.5	15,883.9	14,530.4	13,292.1	12,159.3	11,123.0	10,175.1	9,308.0	8,514.9	226,664.1	375
29,630.4	27,105.3	24,795.5	22,682.5	20,749.4	18,981.1	17,363.5	15,883.9	14,530.4	13,292.1	12,159.3	11,123.0	10,175.1	9,308.0	247,779.5	380
32,390.7	29,630.4	27,105.3	24,795.5	22,682.5	20,749.4	18,981.1	17,363.5	15,883.9	14,530.4	13,292.1	12,159.3	11,123.0	10,175.1	270,862.2	385
35,408.2	32,390.7	29,630.4	27,105.3	24,795.5	22,682.5	20,749.4	18,981.1	17,363.5	15,883.9	14,530.4	13,292.1	12,159.3	11,123.0	296,095.4	390
38,706.8	35,408.2	32,390.7	29,630.4	27,105.3	24,795.5	22,682.5	20,749.4	18,981.1	17,363.5	15,883.9	14,530.4	13,292.1	12,159.3	323,679.2	395
42,312.6	38,706.8	35,408.2	32,390.7	29,630.4	27,105.3	24,795.5	22,682.5	20,749.4	18,981.1	17,363.5	15,883.9	14,530.4	13,292.1	353,832.5	400
46,254.4	42,312.6	38,706.8	35,408.2	32,390.7	29,630.4	27,105.3	24,795.5	22,682.5	20,749.4	18,981.1	17,363.5	15,883.9	14,530.4	386,794.7	405
50,563.3	46,254.4	42,312.6	38,706.8	35,408.2	32,390.7	29,630.4	27,105.3	24,795.5	22,682.5	20,749.4	18,981.1	17,363.5	15,883.9	422,827.7	410
55,273.7	50,563.3	46,254.4	42,312.6	38,706.8	35,408.2	32,390.7	29,630.4	27,105.3	24,795.5	22,682.5	20,749.4	18,981.1	17,363.5	462,217.5	415
60,422.9	55,273.7	50,563.3	46,254.4	42,312.6	38,706.8	35,408.2	32,390.7	29,630.4	27,105.3	24,795.5	22,682.5	20,749.4	18,981.1	505,276.9	420
66,051.8	60,422.9	55,273.7	50,563.3	46,254.4	42,312.6	38,706.8	35,408.2	32,390.7	29,630.4	27,105.3	24,795.5	22,682.5	20,749.4	552,347.7	425
72,205.1	66,051.8	60,422.9	55,273.7	50,563.3	46,254.4	42,312.6	38,706.8	35,408.2	32,390.7	29,630.4	27,105.3	24,795.5	22,682.5	603,803.3	430

Design	Starting Population	Ending Population	Life Expectancy	Fractional	Infant Mortality	Birth Rate Required
001	70	603,550	50	Equal	No	3.584
002	66	603,550	50	Equal	No	3.598
009	140	625,550	50	Equal	No	3.438
046	132	625,550	70	Equal	No	3.487
113	376	603,550	70	Unequal	No	3.366
114	66 M only	603 M/W/C	70	Unequal	No	
235	70 M/W/C	625 M no W	50	Unequal	Yes	
236	66 M/W/C	625 M no W	50	Unequal	Yes	
159	70 M no W	603 M no W	50	Equal	Yes	
196	66 M no W	603 M no W	70	Equal	Yes	
203	70 M only	625 M no W	70	Equal	Yes	
204	66 M only	625 M no W	70	Equal	Yes	
097	70 M/W/C	603 M only	50	Unequal	No	
098	66 M/W/C	603 M only	50	Unequal	No	
105	70 M no W	625 M only	50	Unequal	No	
142	66 M no W	625 M only	70	Unequal	No	
065	70 M only	603 M only	70	Equal	No	
066	66 M only	603 M only	70	Equal	No	

Birth rate per (2) persons per (25) years: 3.366
Birth rate per (1) person per (25) years: 1.683
Birth rate per (1) person per (5) years: 0.337

Total starting population of 376 persons.
Total ending population 603,550 persons.
Israel stayed in Egypt 430 years.

Numerical analysis 113

0 to 5 Yrs	5 to 10 Yrs	10 to 15 Yrs	15 to 20 Yrs	20 to 25 Yrs	25 to 30 Yrs	30 to 35 Yrs	35 to 40 Yrs	40 to 45 Yrs	45 to 50 Yrs	50 to 55 Yrs	55 to 60 Yrs	60 to 65 Yrs	65 to 70 Yrs	Total Population	Years
			0.125	0.270	0.286	0.222	0.096								
26.86	26.86	26.86	26.86	26.86	26.86	26.86	26.86	26.86	26.86	26.86	26.86	26.86	26.86	376.0	0
45.2	26.9	26.9	26.9	26.9	26.9	26.9	26.9	26.9	26.9	26.9	26.9	26.9	26.9	394.3	5
45.2	45.2	26.9	26.9	26.9	26.9	26.9	26.9	26.9	26.9	26.9	26.9	26.9	26.9	412.6	10
45.2	45.2	45.2	26.9	26.9	26.9	26.9	26.9	26.9	26.9	26.9	26.9	26.9	26.9	430.9	15
49.0	45.2	45.2	45.2	26.9	26.9	26.9	26.9	26.9	26.9	26.9	26.9	26.9	26.9	449.2	20
57.3	49.0	45.2	45.2	45.2	26.9	26.9	26.9	26.9	26.9	26.9	26.9	26.9	26.9	471.4	25
66.1	57.3	49.0	45.2	45.2	45.2	26.9	26.9	26.9	26.9	26.9	26.9	26.9	26.9	501.9	30
73.0	66.1	57.3	49.0	45.2	45.2	45.2	26.9	26.9	26.9	26.9	26.9	26.9	26.9	541.1	35
76.7	73.0	66.1	57.3	49.0	45.2	45.2	45.2	26.9	26.9	26.9	26.9	26.9	26.9	587.2	40
80.2	76.7	73.0	66.1	57.3	49.0	45.2	45.2	45.2	26.9	26.9	26.9	26.9	26.9	637.1	45
87.7	80.2	76.7	73.0	66.1	57.3	49.0	45.2	45.2	45.2	26.9	26.9	26.9	26.9	690.5	50
98.6	87.7	80.2	76.7	73.0	66.1	57.3	49.0	45.2	45.2	45.2	26.9	26.9	26.9	751.4	55
110.5	98.6	87.7	80.2	76.7	73.0	66.1	57.3	49.0	45.2	45.2	45.2	26.9	26.9	823.1	60
120.8	110.5	98.6	87.7	80.2	76.7	73.0	66.1	57.3	49.0	45.2	45.2	45.2	26.9	906.7	65
129.8	120.8	110.5	98.6	87.7	80.2	76.7	73.0	66.1	57.3	49.0	45.2	45.2	45.2	1,000.7	70
139.7	129.8	120.8	110.5	98.6	87.7	80.2	76.7	73.0	66.1	57.3	49.0	45.2	45.2	1,085.3	75
152.6	139.7	129.8	120.8	110.5	98.6	87.7	80.2	76.7	73.0	66.1	57.3	49.0	45.2	1,179.9	80
168.8	152.6	139.7	129.8	120.8	110.5	98.6	87.7	80.2	76.7	73.0	66.1	57.3	49.0	1,287.4	85
186.4	168.8	152.6	139.7	129.8	120.8	110.5	98.6	87.7	80.2	76.7	73.0	66.1	57.3	1,411.0	90
203.7	186.4	168.8	152.6	139.7	129.8	120.8	110.5	98.6	87.7	80.2	76.7	73.0	66.1	1,548.4	95
221.1	203.7	186.4	168.8	152.6	139.7	129.8	120.8	110.5	98.6	87.7	80.2	76.7	73.0	1,694.8	100
	221.1	203.7	186.4	168.8	152.6	139.7	129.8	120.8	110.5	98.6	87.7	80.2	76.7	1,849.8	105

Sample Calculations:

Calculating new children in Year 85 between 0 to 5 Yrs of age.

	Persons	Fractional	Multiplier	Calculation
Persons in Year 80 between 15 to 20 Years	110.5	0.125	5	69.05
Persons in Year 80 between 20 to 25 Years	98.6	0.270	5	133.12
Persons in Year 80 between 25 to 30 Years	87.7	0.286	5	125.44
Persons in Year 80 between 30 to 35 Years	80.2	0.222	5	89.06
Persons in Year 80 between 35 to 40 Years	76.7	0.096	5	36.83
				453.50

Birth rate per (1) person per (5) years: 0.337

New children in Year 85 between 0 to 5 Yrs of age. 152.6

Machine Design Analysis 114

Starting Population

	66

Starting Population

		M only

Ending Population

603	

Ending Population

M/W/C		

Life Expectancy

	70

Fractional

	Unequal

Infant Mortality

No	

Design	Starting Population	Ending Population	Life Expectancy	Fractional	Infant Mortality	Birth Rate Required
001	70	603,550	50	Equal	No	3.584
002	66	603,550	50	Equal	No	3.598
009	140	625,550	50	Equal	No	3.438
046	132	625,550	70	Equal	No	3.487
113	376	603,550	70	Unequal	No	3.366
114	355	603,550	70	Unequal	No	3.380
235	70 M/W/C	625 M no W	50	Unequal	Yes	
236	66 M/W/C	625 M no W	50	Unequal	Yes	
159	70 M no W	603 M no W	50	Equal	Yes	
196	66 M no W	603 M no W	70	Equal	Yes	
203	70 M only	625 M no W	70	Equal	Yes	
204	66 M only	625 M no W	70	Equal	Yes	
097	70 M/W/C	603 M only	50	Unequal	No	
098	66 M/W/C	603 M only	50	Unequal	No	
105	70 M no W	625 M only	50	Unequal	No	
142	66 M no W	625 M only	70	Unequal	No	
065	70 M only	603 M only	70	Equal	No	
066	66 M only	603 M only	70	Equal	No	

Machine Design Analysis 235

Starting Population

| 70 | | |

Starting Population

| M/W/C | | |

Ending Population

| | 625 |

Ending Population

| | M no W | |

Life Expectancy

| 50 | |

Fractional

| | Unequal |

Infant Mortality

| | Yes |

Design	Starting Population	Ending Population	Life Expectancy	Fractional	Infant Mortality	Birth Rate Required
001	70	603,550	50	Equal	No	3.584
002	66	603,550	50	Equal	No	3.598
009	140	625,550	50	Equal	No	3.438
046	132	625,550	70	Equal	No	3.487
113	376	603,550	70	Unequal	No	3.366
114	355	603,550	70	Unequal	No	3.380
235	70	1,251,100	50	Unequal	Yes	4.235
236	66 M/W/C	625 M no W	50	Unequal	Yes	
159	70 M no W	603 M no W	50	Equal	Yes	
196	66 M no W	603 M no W	70	Equal	Yes	
203	70 M only	625 M no W	70	Equal	Yes	
204	66 M only	625 M no W	70	Equal	Yes	
097	70 M/W/C	603 M only	50	Unequal	No	
098	66 M/W/C	603 M only	50	Unequal	No	
105	70 M no W	625 M only	50	Unequal	No	
142	66 M no W	625 M only	70	Unequal	No	
065	70 M only	603 M only	70	Equal	No	
066	66 M only	603 M only	70	Equal	No	

Machine Design Analysis 236

Starting Population

	66

Starting Population

M/W/C		

Ending Population

	625

Ending Population

	M no W	

Life Expectancy

50	

Fractional

	Unequal

Infant Mortality

	Yes

Design	Starting Population	Ending Population	Life Expectancy	Fractional	Infant Mortality	Birth Rate Required
001	70	603,550	50	Equal	No	3.584
002	66	603,550	50	Equal	No	3.598
009	140	625,550	50	Equal	No	3.438
046	132	625,550	70	Equal	No	3.487
113	376	603,550	70	Unequal	No	3.366
114	355	603,550	70	Unequal	No	3.380
235	70	1,251,100	50	Unequal	Yes	4.235
236	66	1,251,100	50	Unequal	Yes	4.252
159	70 M no W	603 M no W	50	Equal	Yes	
196	66 M no W	603 M no W	70	Equal	Yes	
203	70 M only	625 M no W	70	Equal	Yes	
204	66 M only	625 M no W	70	Equal	Yes	
097	70 M/W/C	603 M only	50	Unequal	No	
098	66 M/W/C	603 M only	50	Unequal	No	
105	70 M no W	625 M only	50	Unequal	No	
142	66 M no W	625 M only	70	Unequal	No	
065	70 M only	603 M only	70	Equal	No	
066	66 M only	603 M only	70	Equal	No	

Question: Why did the author skip the details for machine design analysis 114, 235, and 236?

Answer: The details for machine design analysis 114, 235, and 236 are similar to the previous machine design analysis. The author will show details for selected machine design analysis and the remaining machine design analysis will have summary data only.

Question: How does the author account for miscarriages and other unfortunate life events in his numerical analysis of Israel population growth from Genesis to Exodus?

Answer 1: The author uses five year time integration which inherently accounts for (6) live births, (2) miscarriages, and (2) infant mortalities per woman per twenty years. See the next sheet for details.

	Months	Accumulated Months	Live Births	Miscarriages	Infant Mortality
Trying to get pregnant	4	4			
Nine months of pregnancy	9	13	1		
Nursing and waiting period before trying to get pregnant again	6	19			
Trying to get pregnant	4	23			
Five months of pregnancy, then miscarriage	5	28		1	
Grieving and waiting period before trying to get pregnant again	12	40			
Trying to get pregnant	4	44			
Nine months of pregnancy	9	53	1		
Nursing and waiting period before trying to get pregnant again	6	59			
Trying to get pregnant	4	63			
Nine months of pregnancy, then sudden infant death at 12 months	21	84			1
Grieving and waiting period before trying to get pregnant again	12	96			
Trying to get pregnant	4	100			
Nine months of pregnancy	9	109	1		
Nursing and waiting period before trying to get pregnant again	6	115			
In ten years, the 5 year time integration accommodates (3) live births, (1) miscarriage, and (1) infant mortality per woman					
In twenty years, the 5 year time integration accommodates (6) live births, (2) miscarriage, and (2) infant mortality per woman					

Answer 2: The author could mathematically adjust for miscarriages and other unfortunate life events in his calculations of Israel population growth from Genesis to Exodus.

This book uses infant mortality (No) and infant mortality (Yes) as the two possibilities when analyzing Israel population growth from Genesis to Exodus.

Infant mortality (No) equals five year time integration only.

Infant mortality (Yes) equals five year time integration plus mathematical adjustment

See sample calculation after machine design analysis 159.

Machine Design Analysis 159

Starting Population

| 70 | |

Starting Population

| | M no W | |

Ending Population

| 603 | |

Ending Population

| | M no W | |

Life Expectancy

| 50 | |

Fractional

| Equal | |

Infant Mortality

| | Yes |

For the starting population (Genesis 46: 27) assume the word "men" means males only and we need to account for females (70 M no W); therefore, the total starting population would be (70 * 2) equals 140 persons.

Assume the time the Israelites had stayed in Egypt was four hundred and thirty years (Exodus 12:40).

For the ending population (Numbers 1: 1-3, Numbers 1: 44-46) assume the word "men" means males only and we need to account for females (603 M no W); therefore, the total ending population would be (603,550 * 2) equals 1,207,100 persons.

Assume a human life expectancy of 50 years. Using numerical analysis to calculate the birth rate required for a total starting population of 140 persons to increase to a total ending population of 1,207,100 persons in 430 years results in a birth rate of 3.868 children.

Numerical analysis 159

Birth rate per (2) persons per (20) years	3.868
Birth rate per (1) person per (20) years	1.934
Birth rate per (1) person per (5) years	0.484
Infant Mortality Rate (per 1000)	73.30

Total starting population of 140 persons.
Total ending population of 1,207,100 persons
Israel stayed in Egypt 430 years.

0 to 5 Yrs	5 to 10 Yrs	10 to 15 Yrs	15 to 20 Yrs	20 to 25 Yrs	25 to 30 Yrs	30 to 35 Yrs	35 to 40 Yrs	40 to 45 Yrs	45 to 50 Yrs	Total Population	Years
14.0	14.0	14.0	14.0	14.0	14.0	14.0	14.0	14.0	14.0	140.0	0
25.1	14.0	14.0	14.0	14.0	14.0	14.0	14.0	14.0	14.0	151.1	5
25.1	25.1	14.0	14.0	14.0	14.0	14.0	14.0	14.0	14.0	162.2	10
25.1	25.1	25.1	14.0	14.0	14.0	14.0	14.0	14.0	14.0	173.3	15
25.1	25.1	25.1	25.1	14.0	14.0	14.0	14.0	14.0	14.0	184.4	20
30.1	25.1	25.1	25.1	25.1	14.0	14.0	14.0	14.0	14.0	200.4	25
35.0	30.1	25.1	25.1	25.1	25.1	14.0	14.0	14.0	14.0	221.5	30
40.0	35.0	30.1	25.1	25.1	25.1	25.1	14.0	14.0	14.0	247.5	35
45.0	40.0	35.0	30.1	25.1	25.1	25.1	25.1	14.0	14.0	278.4	40
47.2	45.0	40.0	35.0	30.1	25.1	25.1	25.1	25.1	14.0	311.6	45
51.6	47.2	45.0	40.0	35.0	30.1	25.1	25.1	25.1	25.1	349.3	50
58.3	51.6	47.2	45.0	40.0	35.0	30.1	25.1	25.1	25.1	382.5	55
67.2	58.3	51.6	47.2	45.0	40.0	35.0	30.1	25.1	25.1	424.7	60
74.9	67.2	58.3	51.6	47.2	45.0	40.0	35.0	30.1	25.1	474.5	65
82.4	74.9	67.2	58.3	51.6	47.2	45.0	40.0	35.0	30.1	531.7	70
90.6	82.4	74.9	67.2	58.3	51.6	47.2	45.0	40.0	35.0	592.3	75
100.5	90.6	82.4	74.9	67.2	58.3	51.6	47.2	45.0	40.0	657.8	80
113.0	100.5	90.6	82.4	74.9	67.2	58.3	51.6	47.2	45.0	730.7	85
126.7	113.0	100.5	90.6	82.4	74.9	67.2	58.3	51.6	47.2	812.5	90
141.2	126.7	113.0	100.5	90.6	82.4	74.9	67.2	58.3	51.6	906.5	95
156.1	141.2	126.7	113.0	100.5	90.6	82.4	74.9	67.2	58.3	1,010.9	100
173.2	156.1	141.2	126.7	113.0	100.5	90.6	82.4	74.9	67.2	1,125.8	105
193.0	173.2	156.1	141.2	126.7	113.0	100.5	90.6	82.4	74.9	1,251.6	110
215.7	193.0	173.2	156.1	141.2	126.7	113.0	100.5	90.6	82.4	1,392.4	115
240.6	215.7	193.0	173.2	156.1	141.2	126.7	113.0	100.5	90.6	1,550.6	120
267.6	240.6	215.7	193.0	173.2	156.1	141.2	126.7	113.0	100.5	1,727.6	125
297.3	267.6	240.6	215.7	193.0	173.2	156.1	141.2	126.7	113.0	1,924.3	130
330.7	297.3	267.6	240.6	215.7	193.0	173.2	156.1	141.2	126.7	2,142.0	135
368.5	330.7	297.3	267.6	240.6	215.7	193.0	173.2	156.1	141.2	2,383.8	140
410.8	368.5	330.7	297.3	267.6	240.6	215.7	193.0	173.2	156.1	2,653.4	145
457.5	410.8	368.5	330.7	297.3	267.6	240.6	215.7	193.0	173.2	2,954.9	150
509.0	457.5	410.8	368.5	330.7	297.3	267.6	240.6	215.7	193.0	3,290.8	155
566.4	509.0	457.5	410.8	368.5	330.7	297.3	267.6	240.6	215.7	3,664.1	160
630.5	566.4	509.0	457.5	410.8	368.5	330.7	297.3	267.6	240.6	4,078.9	165
702.4	630.5	566.4	509.0	457.5	410.8	368.5	330.7	297.3	267.6	4,540.7	170
782.3	702.4	630.5	566.4	509.0	457.5	410.8	368.5	330.7	297.3	5,055.4	175
870.9	782.3	702.4	630.5	566.4	509.0	457.5	410.8	368.5	330.7	5,629.1	180
969.4	870.9	782.3	702.4	630.5	566.4	509.0	457.5	410.8	368.5	6,267.8	185
1,079.1	969.4	870.9	782.3	702.4	630.5	566.4	509.0	457.5	410.8	6,978.3	190
1,201.5	1,079.1	969.4	870.9	782.3	702.4	630.5	566.4	509.0	457.5	7,769.0	195
1,338.0	1,201.5	1,079.1	969.4	870.9	782.3	702.4	630.5	566.4	509.0	8,649.4	200
1,489.8	1,338.0	1,201.5	1,079.1	969.4	870.9	782.3	702.4	630.5	566.4	9,630.1	205

0 to 5 Yrs	5 to 10 Yrs	10 to 15 Yrs	15 to 20 Yrs	20 to 25 Yrs	25 to 30 Yrs	30 to 35 Yrs	35 to 40 Yrs	40 to 45 Yrs	45 to 50 Yrs	Total Population	Years
1,658.5	1,489.8	1,338.0	1,201.5	1,079.1	969.4	870.9	782.3	702.4	630.5	10,722.3	210
1,846.4	1,658.5	1,489.8	1,338.0	1,201.5	1,079.1	969.4	870.9	782.3	702.3	11,938.1	215
2,055.6	1,846.4	1,658.5	1,489.8	1,338.0	1,201.5	1,079.1	969.4	870.9	782.3	13,291.4	220
2,288.8	2,055.6	1,846.4	1,658.5	1,489.8	1,338.0	1,201.5	1,079.1	969.4	870.9	14,797.9	225
2,548.5	2,288.8	2,055.6	1,846.4	1,658.5	1,489.8	1,338.0	1,201.5	1,079.1	969.4	16,475.5	230
2,837.4	2,548.5	2,288.8	2,055.6	1,846.4	1,658.5	1,489.8	1,338.0	1,201.5	1,079.1	18,343.5	235
3,159.0	2,837.4	2,548.5	2,288.8	2,055.6	1,846.4	1,658.5	1,489.8	1,338.0	1,201.5	20,423.4	240
3,517.0	3,159.0	2,837.4	2,548.5	2,288.8	2,055.6	1,846.4	1,658.5	1,489.8	1,338.0	22,738.9	245
3,915.7	3,517.0	3,159.0	2,837.4	2,548.5	2,288.8	2,055.6	1,846.4	1,658.5	1,489.8	25,316.7	250
4,359.8	3,915.7	3,517.0	3,159.0	2,837.4	2,548.5	2,288.8	2,055.6	1,846.4	1,658.5	28,186.7	255
4,854.1	4,359.8	3,915.7	3,517.0	3,159.0	2,837.4	2,548.5	2,288.8	2,055.6	1,846.4	31,382.2	260
5,404.4	4,854.1	4,359.8	3,915.7	3,517.0	3,159.0	2,837.4	2,548.5	2,288.8	2,055.6	34,940.2	265
6,017.0	5,404.4	4,854.1	4,359.8	3,915.7	3,517.0	3,159.0	2,837.4	2,548.5	2,288.8	38,901.6	270
6,699.1	6,017.0	5,404.4	4,854.1	4,359.8	3,915.7	3,517.0	3,159.0	2,837.4	2,548.5	43,311.9	275
7,458.7	6,699.1	6,017.0	5,404.4	4,854.1	4,359.8	3,915.7	3,517.0	3,159.0	2,837.4	48,222.1	280
8,304.3	7,458.7	6,699.1	6,017.0	5,404.4	4,854.1	4,359.8	3,915.7	3,517.0	3,159.0	53,689.1	285
9,245.8	8,304.3	7,458.7	6,699.1	6,017.0	5,404.4	4,854.1	4,359.8	3,915.7	3,517.0	59,775.9	290
10,294.0	9,245.8	8,304.3	7,458.7	6,699.1	6,017.0	5,404.4	4,854.1	4,359.8	3,915.7	66,553.0	295
11,461.0	10,294.0	9,245.8	8,304.3	7,458.7	6,699.1	6,017.0	5,404.4	4,854.1	4,359.8	74,098.3	300
12,760.3	11,461.0	10,294.0	9,245.8	8,304.3	7,458.7	6,699.1	6,017.0	5,404.4	4,854.1	82,498.8	305
14,207.0	12,760.3	11,461.0	10,294.0	9,245.8	8,304.3	7,458.7	6,699.1	6,017.0	5,404.4	91,851.8	310
15,817.8	14,207.0	12,760.3	11,461.0	10,294.0	9,245.8	8,304.3	7,458.7	6,699.1	6,017.0	102,265.1	315
17,611.1	15,817.8	14,207.0	12,760.3	11,461.0	10,294.0	9,245.8	8,304.3	7,458.7	6,699.1	113,859.2	320
19,607.6	17,611.1	15,817.8	14,207.0	12,760.3	11,461.0	10,294.0	9,245.8	8,304.3	7,458.7	126,767.7	325
21,830.5	19,607.6	17,611.1	15,817.8	14,207.0	12,760.3	11,461.0	10,294.0	9,245.8	8,304.3	141,139.6	330
24,305.5	21,830.5	19,607.6	17,611.1	15,817.8	14,207.0	12,760.3	11,461.0	10,294.0	9,245.8	157,140.7	335
27,061.1	24,305.5	21,830.5	19,607.6	17,611.1	15,817.8	14,207.0	12,760.3	11,461.0	10,294.0	174,956.0	340
30,129.1	27,061.1	24,305.5	21,830.5	19,607.6	17,611.1	15,817.8	14,207.0	12,760.3	11,461.0	194,791.1	345
33,544.9	30,129.1	27,061.1	24,305.5	21,830.5	19,607.6	17,611.1	15,817.8	14,207.0	12,760.3	216,874.9	350
37,347.9	33,544.9	30,129.1	27,061.1	24,305.5	21,830.5	19,607.6	17,611.1	15,817.8	14,207.0	241,462.4	355
41,582.0	37,347.9	33,544.9	30,129.1	27,061.1	24,305.5	21,830.5	19,607.6	17,611.1	15,817.8	268,837.4	360
46,296.3	41,582.0	37,347.9	33,544.9	30,129.1	27,061.1	24,305.5	21,830.5	19,607.6	17,611.1	299,316.0	365
51,545.0	46,296.3	41,582.0	37,347.9	33,544.9	30,129.1	27,061.1	24,305.5	21,830.5	19,607.6	333,249.9	370
57,388.8	51,545.0	46,296.3	41,582.0	37,347.9	33,544.9	30,129.1	27,061.1	24,305.5	21,830.5	371,031.0	375
63,895.0	57,388.8	51,545.0	46,296.3	41,582.0	37,347.9	33,544.9	30,129.1	27,061.1	24,305.5	413,095.5	380
71,138.9	63,895.0	57,388.8	51,545.0	46,296.3	41,582.0	37,347.9	33,544.9	30,129.1	27,061.1	459,928.9	385
79,204.0	71,138.9	63,895.0	57,388.8	51,545.0	46,296.3	41,582.0	37,347.9	33,544.9	30,129.1	512,071.8	390
88,183.5	79,204.0	71,138.9	63,895.0	57,388.8	51,545.0	46,296.3	41,582.0	37,347.9	33,544.9	570,126.2	395
98,181.0	88,183.5	79,204.0	71,138.9	63,895.0	57,388.8	51,545.0	46,296.3	41,582.0	37,347.9	634,762.4	400
109,312.0	98,181.0	88,183.5	79,204.0	71,138.9	63,895.0	57,388.8	51,545.0	46,296.3	41,582.0	706,726.5	405
121,704.9	109,312.0	98,181.0	88,183.5	79,204.0	71,138.9	63,895.0	57,388.8	51,545.0	46,296.3	786,849.4	410
135,502.8	121,704.9	109,312.0	98,181.0	88,183.5	79,204.0	71,138.9	63,895.0	57,388.8	51,545.0	876,055.9	415
150,865.0	135,502.8	121,704.9	109,312.0	98,181.0	88,183.5	79,204.0	71,138.9	63,895.0	57,388.8	975,375.8	420
167,968.8	150,865.0	135,502.8	121,704.9	109,312.0	98,181.0	88,183.5	79,204.0	71,138.9	63,895.0	1,085,955.8	425
187,011.7	167,968.8	150,865.0	135,502.8	121,704.9	109,312.0	98,181.0	88,183.5	79,204.0	71,138.9	1,209,072.5	430

Design	Starting Population	Ending Population	Life Expectancy	Fractional	Infant Mortality	Birth Rate Required
001	70	603,550	50	Equal	No	3.584
002	66	603,550	50	Equal	No	3.598
009	140	625,550	50	Equal	No	3.438
046	132	625,550	70	Equal	No	3.487
113	376	603,550	70	Unequal	No	3.366
114	355	603,550	70	Unequal	No	3.380
235	70	1,251,100	50	Unequal	Yes	4.235
236	66	1,251,100	50	Unequal	Yes	4.252
159	140	1,207,100	50	Equal	Yes	3.868
196	66 M no W	603 M no W	70	Equal	Yes	
203	70 M only	625 M no W	70	Equal	Yes	
204	66 M only	625 M no W	70	Equal	Yes	
097	70 M/W/C	603 M only	50	Unequal	No	
098	66 M/W/C	603 M only	50	Unequal	No	
105	70 M no W	625 M only	50	Unequal	No	
142	66 M no W	625 M only	70	Unequal	No	
065	70 M only	603 M only	70	Equal	No	
066	66 M only	603 M only	70	Equal	No	

| | | | | | | | | | | Numerical analysis 159 | |

Birth rate per (2) persons per (20) years 3.868
Birth rate per (1) person per (20) years 1.934
Birth rate per (1) person per (5) years 0.484
Infant Mortality Rate (per 1000) 73.30

Total starting population of 140 persons.
Total ending population of 1,207,100 persons.
Israel stayed in Egypt 430 years.

0 to 5 Yrs	5 to 10 Yrs	10 to 15 Yrs	15 to 20 Yrs	20 to 25 Yrs	25 to 30 Yrs	30 to 35 Yrs	35 to 40 Yrs	40 to 45 Yrs	45 to 50 Yrs	Total Population	Years
14.0	14.0	14.0	14.0	14.0	14.0	14.0	14.0	14.0	14.0	140.0	0
25.1	14.0	14.0	14.0	14.0	14.0	14.0	14.0	14.0	14.0	151.1	5
25.1	25.1	14.0	14.0	14.0	14.0	14.0	14.0	14.0	14.0	162.2	10
25.1	25.1	25.1	14.0	14.0	14.0	14.0	14.0	14.0	14.0	173.3	15
25.1	25.1	25.1	25.1	14.0	14.0	14.0	14.0	14.0	14.0	184.4	20
30.1	25.1	25.1	25.1	25.1	14.0	14.0	14.0	14.0	14.0	200.4	25
35.0	30.1	25.1	25.1	25.1	25.1	14.0	14.0	14.0	14.0	221.5	30
40.0	35.0	30.1	25.1	25.1	25.1	25.1	14.0	14.0	14.0	247.5	35
45.0	40.0	35.0	30.1	25.1	25.1	25.1	25.1	14.0	14.0	278.4	40
47.2	45.0	40.0	35.0	30.1	25.1	25.1	25.1	25.1	14.0	311.6	45
51.6	47.2	45.0	40.0	35.0	30.1	25.1	25.1	25.1	25.1	349.3	50
58.3	51.6	47.2	45.0	40.0	35.0	30.1	25.1	25.1	25.1	382.5	55
67.2	58.3	51.6	47.2	45.0	40.0	35.0	30.1	25.1	25.1	424.7	60
74.9	67.2	58.3	51.6	47.2	45.0	40.0	35.0	30.1	25.1	474.5	65
82.4	74.9	67.2	58.3	51.6	47.2	45.0	40.0	35.0	30.1	531.7	70
90.6	82.4	74.9	67.2	58.3	51.6	47.2	45.0	40.0	35.0	592.3	75
100.5	90.6	82.4	74.9	67.2	58.3	51.6	47.2	45.0	40.0	657.8	80
113.0	100.5	90.6	82.4	74.9	67.2	58.3	51.6	47.2	45.0	730.7	85
126.7	113.0	100.5	90.6	82.4	74.9	67.2	58.3	51.6	47.2	812.5	90
141.2	126.7	113.0	100.5	90.6	82.4	74.9	67.2	58.3	51.6	906.5	95

Sample Calculations:

Calculating New Children in Year 85 between 0 to 5 Yrs of age.

Persons in Year 80 between 15 to 20 Years	74.9
Persons in Year 80 between 20 to 25 Years	67.2
Persons in Year 80 between 25 to 30 Years	58.3
Persons in Year 80 between 30 to 35 Years	51.6
Total persons	252.1
Birth Rate per person (1) person per (5) years	0.484
Infant Mortality Rate (per 1000)	73.30
New children in Year 85 between 0 to 5 Yrs of age.	113.0

Question: How did the author determine the infant mortality rate?

Answer: The author applied statistical analysis to the infant mortality rates specified in *The World Almanac and Book of Facts 2007* (World Almanac Books, ISBN # 978-0-88687-995-2) for selected countries.

The countries selected are based on their close geographic proximity to the lands associated with the bible. Please see the following page for details.

Country	Infant Mortality Rate per 1000 Births
Afghanistan	160.2
Egypt	31.3
Iran	39.3
Iraq	48.6
Israel	6.9
Jordan	16.8
Kuwait	9.7
Lebanon	23.7
Oman	18.9
Pakistan	70.5
Qatar	18.0
Saudi Arabia	12.8
Syria	28.6
Turkey	39.7
UAE	14.1
Yemen	59.9
Minimum	6.9
Maximum	160.2
Average	37.4
Standard Dev	36.3
Avg. + 1 Std	73.7

Machine Design Analysis 196

Starting Population

	66

Starting Population
M no W

Ending Population
603

Ending Population
M no W

Life Expectancy

	70

Fractional
Equal

Infant Mortality

	Yes

Design	Starting Population	Ending Population	Life Expectancy	Fractional	Infant Mortality	Birth Rate Required
001	70	603,550	50	Equal	No	3.584
002	66	603,550	50	Equal	No	3.598
009	140	625,550	50	Equal	No	3.438
046	132	625,550	70	Equal	No	3.487
113	376	603,550	70	Unequal	No	3.366
114	355	603,550	70	Unequal	No	3.380
235	70	1,251,100	50	Unequal	Yes	4.235
236	66	1,251,100	50	Unequal	Yes	4.252
159	140	1,207,100	50	Equal	Yes	3.868
196	132	1,207,100	70	Equal	Yes	3.925
203	70 M only	625 M no W	70	Equal	Yes	
204	66 M only	625 M no W	70	Equal	Yes	
097	70 M/W/C	603 M only	50	Unequal	No	
098	66 M/W/C	603 M only	50	Unequal	No	
105	70 M no W	625 M only	50	Unequal	No	
142	66 M no W	625 M only	70	Unequal	No	
065	70 M only	603 M only	70	Equal	No	
066	66 M only	603 M only	70	Equal	No	

Machine Design Analysis 203

Starting Population

70		

Starting Population

		M only

Ending Population

	625

Ending Population

	M no W	

Life Expectancy

	70

Fractional

Equal	

Infant Mortality

	Yes

Design	Starting Population	Ending Population	Life Expectancy	Fractional	Infant Mortality	Birth Rate Required
001	70	603,550	50	Equal	No	3.584
002	66	603,550	50	Equal	No	3.598
009	140	625,550	50	Equal	No	3.438
046	132	625,550	70	Equal	No	3.487
113	376	603,550	70	Unequal	No	3.366
114	355	603,550	70	Unequal	No	3.380
235	70	1,251,100	50	Unequal	Yes	4.235
236	66	1,251,100	50	Unequal	Yes	4.252
159	140	1,207,100	50	Equal	Yes	3.868
196	132	1,207,100	70	Equal	Yes	3.925
203	397	1,251,100	70	Equal	Yes	3.666
204	66 M only	625 M no W	70	Equal	Yes	
097	70 M/W/C	603 M only	50	Unequal	No	
098	66 M/W/C	603 M only	50	Unequal	No	
105	70 M no W	625 M only	50	Unequal	No	
142	66 M no W	625 M only	70	Unequal	No	
065	70 M only	603 M only	70	Equal	No	
066	66 M only	603 M only	70	Equal	No	

Machine Design Analysis 204

Starting Population

| | 66 |

Starting Population

| | | M only |

Ending Population

| | 625 |

Ending Population

| | M no W | |

Life Expectancy

| | 70 |

Fractional

| Equal | |

Infant Mortality

| | Yes |

Design	Starting Population	Ending Population	Life Expectancy	Fractional	Infant Mortality	Birth Rate Required
001	70	603,550	50	Equal	No	3.584
002	66	603,550	50	Equal	No	3.598
009	140	625,550	50	Equal	No	3.438
046	132	625,550	70	Equal	No	3.487
113	376	603,550	70	Unequal	No	3.366
114	355	603,550	70	Unequal	No	3.380
235	70	1,251,100	50	Unequal	Yes	4.235
236	66	1,251,100	50	Unequal	Yes	4.252
159	140	1,207,100	50	Equal	Yes	3.868
196	132	1,207,100	70	Equal	Yes	3.925
203	397	1,251,100	70	Equal	Yes	3.666
204	375	1,251,100	70	Equal	Yes	3.679
097	70 M/W/C	603 M only	50	Unequal	No	
098	66 M/W/C	603 M only	50	Unequal	No	
105	70 M no W	625 M only	50	Unequal	No	
142	66 M no W	625 M only	70	Unequal	No	
065	70 M only	603 M only	70	Equal	No	
066	66 M only	603 M only	70	Equal	No	

73

Machine Design Analysis 097

Starting Population

70	

Starting Population

M/W/C		

Ending Population

603	

Ending Population

		M only

Life Expectancy

50	

Fractional

	Unequal

Infant Mortality

No	

Design	Starting Population	Ending Population	Life Expectancy	Fractional	Infant Mortality	Birth Rate Required
001	70	603,550	50	Equal	No	3.584
002	66	603,550	50	Equal	No	3.598
009	140	625,550	50	Equal	No	3.438
046	132	625,550	70	Equal	No	3.487
113	376	603,550	70	Unequal	No	3.366
114	355	603,550	70	Unequal	No	3.380
235	70	1,251,100	50	Unequal	Yes	4.235
236	66	1,251,100	50	Unequal	Yes	4.252
159	140	1,207,100	50	Equal	Yes	3.868
196	132	1,207,100	70	Equal	Yes	3.925
203	397	1,251,100	70	Equal	Yes	3.666
204	375	1,251,100	70	Equal	Yes	3.679
097	70	3,756,495	50	Unequal	No	4.224
098	66 M/W/C	603 M only	50	Unequal	No	
105	70 M no W	625 M only	50	Unequal	No	
142	66 M no W	625 M only	70	Unequal	No	
065	70 M only	603 M only	70	Equal	No	
066	66 M only	603 M only	70	Equal	No	

Machine Design Analysis 098

Starting Population

	66

Starting Population

M/W/C		

Ending Population

603	

Ending Population

		M only

Life Expectancy

50	

Fractional

	Unequal

Infant Mortality

No	

Design	Starting Population	Ending Population	Life Expectancy	Fractional	Infant Mortality	Birth Rate Required
001	70	603,550	50	Equal	No	3.584
002	66	603,550	50	Equal	No	3.598
009	140	625,550	50	Equal	No	3.438
046	132	625,550	70	Equal	No	3.487
113	376	603,550	70	Unequal	No	3.366
114	355	603,550	70	Unequal	No	3.380
235	70	1,251,100	50	Unequal	Yes	4.235
236	66	1,251,100	50	Unequal	Yes	4.252
159	140	1,207,100	50	Equal	Yes	3.868
196	132	1,207,100	70	Equal	Yes	3.925
203	397	1,251,100	70	Equal	Yes	3.666
204	375	1,251,100	70	Equal	Yes	3.679
097	70	3,756,495	50	Unequal	No	4.224
098	66	3,766,756	50	Unequal	No	4.241
105	70 M no W	625 M only	50	Unequal	No	
142	66 M no W	625 M only	70	Unequal	No	
065	70 M only	603 M only	70	Equal	No	
066	66 M only	603 M only	70	Equal	No	

Machine Design Analysis 105

Starting Population

70	

Starting Population

	M no W	

Ending Population

	625

Ending Population

		M only

Life Expectancy

50	

Fractional

	Unequal

Infant Mortality

No	

Design	Starting Population	Ending Population	Life Expectancy	Fractional	Infant Mortality	Birth Rate Required
001	70	603,550	50	Equal	No	3.584
002	66	603,550	50	Equal	No	3.598
009	140	625,550	50	Equal	No	3.438
046	132	625,550	70	Equal	No	3.487
113	376	603,550	70	Unequal	No	3.366
114	355	603,550	70	Unequal	No	3.380
235	70	1,251,100	50	Unequal	Yes	4.235
236	66	1,251,100	50	Unequal	Yes	4.252
159	140	1,207,100	50	Equal	Yes	3.868
196	132	1,207,100	70	Equal	Yes	3.925
203	397	1,251,100	70	Equal	Yes	3.666
204	375	1,251,100	70	Equal	Yes	3.679
097	70	3,756,495	50	Unequal	No	4.224
098	66	3,766,756	50	Unequal	No	4.241
105	140	3,773,943	50	Unequal	No	4.033
142	66 M no W	625 M only	70	Unequal	No	
065	70 M only	603 M only	70	Equal	No	
066	66 M only	603 M only	70	Equal	No	

Machine Design Analysis 142

Starting Population

| | 66 |

Starting Population

| | M no W | |

Ending Population

| | 625 |

Ending Population

| | | M only |

Life Expectancy

| | 70 |

Fractional

| | Unequal |

Infant Mortality

| No | |

Design	Starting Population	Ending Population	Life Expectancy	Fractional	Infant Mortality	Birth Rate Required
001	70	603,550	50	Equal	No	3.584
002	66	603,550	50	Equal	No	3.598
009	140	625,550	50	Equal	No	3.438
046	132	625,550	70	Equal	No	3.487
113	376	603,550	70	Unequal	No	3.366
114	355	603,550	70	Unequal	No	3.380
235	70	1,251,100	50	Unequal	Yes	4.235
236	66	1,251,100	50	Unequal	Yes	4.252
159	140	1,207,100	50	Equal	Yes	3.868
196	132	1,207,100	70	Equal	Yes	3.925
203	397	1,251,100	70	Equal	Yes	3.666
204	375	1,251,100	70	Equal	Yes	3.679
097	70	3,756,495	50	Unequal	No	4.224
098	66	3,766,756	50	Unequal	No	4.241
105	140	3,773,943	50	Unequal	No	4.033
142	132	3,818,357	70	Unequal	No	4.104
065	70 M only	603 M only	70	Equal	No	
066	66 M only	603 M only	70	Equal	No	

Machine Design Analysis 065

Starting Population

| 70 | | |

Starting Population

| | | M only |

Ending Population

| 603 | | |

Ending Population

| | | M only |

Life Expectancy

| | 70 |

Fractional

| Equal | | |

Infant Mortality

| No | | |

Design	Starting Population	Ending Population	Life Expectancy	Fractional	Infant Mortality	Birth Rate Required
001	70	603,550	50	Equal	No	3.584
002	66	603,550	50	Equal	No	3.598
009	140	625,550	50	Equal	No	3.438
046	132	625,550	70	Equal	No	3.487
113	376	603,550	70	Unequal	No	3.366
114	355	603,550	70	Unequal	No	3.380
235	70	1,251,100	50	Unequal	Yes	4.235
236	66	1,251,100	50	Unequal	Yes	4.252
159	140	1,207,100	50	Equal	Yes	3.868
196	132	1,207,100	70	Equal	Yes	3.925
203	397	1,251,100	70	Equal	Yes	3.666
204	375	1,251,100	70	Equal	Yes	3.679
097	70	3,756,495	50	Unequal	No	4.224
098	66	3,766,756	50	Unequal	No	4.241
105	140	3,773,943	50	Unequal	No	4.033
142	132	3,818,357	70	Unequal	No	4.104
065	394	3,394,365	70	Equal	No	3.624
066	66 M only	603 M only	70	Equal	No	

Machine Design Analysis 066

Starting Population

| | | 66 |

Starting Population

| | | | | M only |

Ending Population

| 603 | | |

Ending Population

| | | | | M only |

Life Expectancy

| | | 70 |

Fractional

| Equal | | |

Infant Mortality

| No | | |

Design	Starting Population	Ending Population	Life Expectancy	Fractional	Infant Mortality	Birth Rate Required
001	70	603,550	50	Equal	No	3.584
002	66	603,550	50	Equal	No	3.598
009	140	625,550	50	Equal	No	3.438
046	132	625,550	70	Equal	No	3.487
113	376	603,550	70	Unequal	No	3.366
114	355	603,550	70	Unequal	No	3.380
235	70	1,251,100	50	Unequal	Yes	4.235
236	66	1,251,100	50	Unequal	Yes	4.252
159	140	1,207,100	50	Equal	Yes	3.868
196	132	1,207,100	70	Equal	Yes	3.925
203	397	1,251,100	70	Equal	Yes	3.666
204	375	1,251,100	70	Equal	Yes	3.679
097	70	3,756,495	50	Unequal	No	4.224
098	66	3,766,756	50	Unequal	No	4.241
105	140	3,773,943	50	Unequal	No	4.033
142	132	3,818,357	70	Unequal	No	4.104
065	394	3,394,365	70	Equal	No	3.624
066	372	3,402,211	70	Equal	No	3.637

Machine Design Analysis Summary:

Machine design analysis requires a birth rate of 3.366 to 4.252 children per woman to obtain Israel population growth from Genesis to Exodus.

The claim that the Exodus of Israel from Egypt is myth based on the singular argument that the birth rate required of children per woman during Israel stay in Egypt is so high (10, 11, 12 children per woman) is mathematically invalid.

Statistical Numerical Analysis

The author will now use statistical numerical analysis to calculate the birth rate of children per woman during Israel stay in Egypt to obtain Israel population growth from Genesis to Exodus. For the machine design analysis the author suggested the following variables (see next page). There are 288 total possible statistical numerical analysis combinations (2 * 3 * 2 * 3 * 2 * 2 *2 equals 288). Since our statistical numerical analysis improves with higher analysis samples the author will calculate all 288 statistical numerical analysis combinations for analyzing Israel population growth from Genesis to Exodus. The methodology for the statistical numerical analysis is the same as the machine design analysis; therefore, the author will show summary data only.

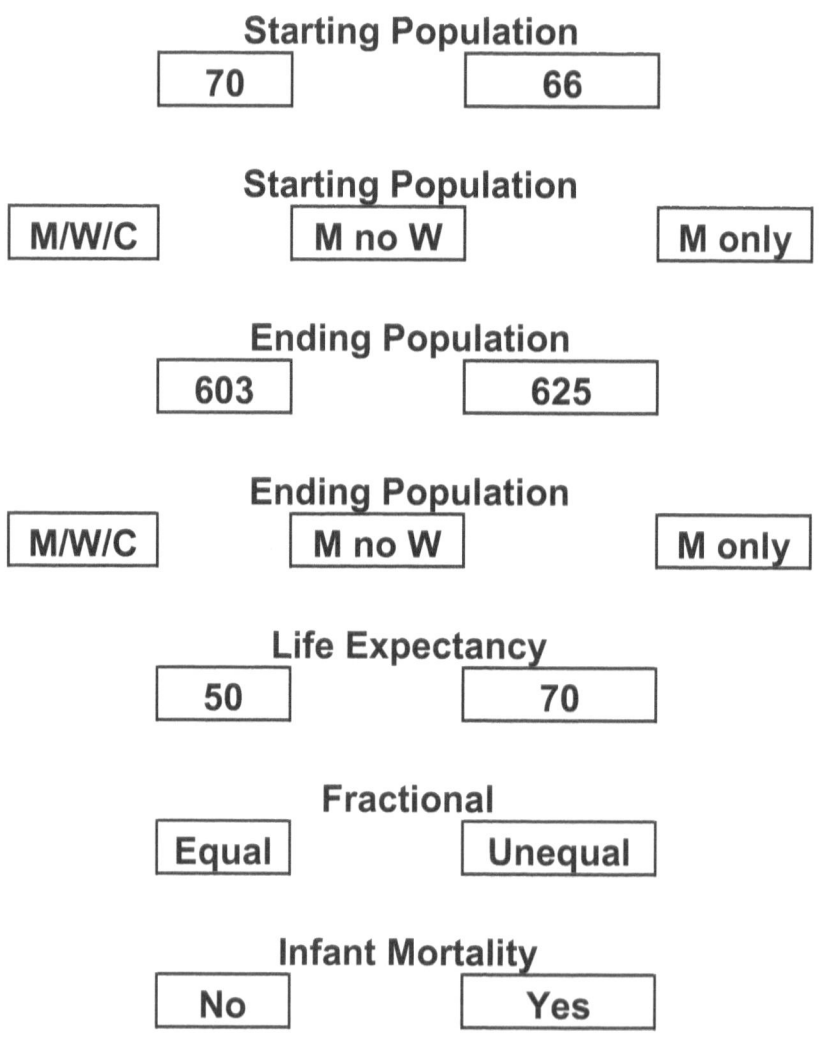

Design	Starting Population	Starting Population	Ending Population	Ending Population	Life Expect.	Fractional	Infant Mortality	Birth Rate Required
1	70	M/W/C	603,550	M/W/C	50	Equal	No	3.584
2	66	M/W/C	603,550	M/W/C	50	Equal	No	3.598
3	70	M no W	603,550	M/W/C	50	Equal	No	3.431
4	66	M no W	603,550	M/W/C	50	Equal	No	3.443
5	70	M only	603,550	M/W/C	50	Equal	No	3.228
6	66	M only	603,550	M/W/C	50	Equal	No	3.239
7	70	M/W/C	625,550	M/W/C	50	Equal	No	3.593
8	66	M/W/C	625,550	M/W/C	50	Equal	No	3.606
9	70	M no W	625,550	M/W/C	50	Equal	No	3.438
10	66	M no W	625,550	M/W/C	50	Equal	No	3.451
11	70	M only	625,550	M/W/C	50	Equal	No	3.235
12	66	M only	625,550	M/W/C	50	Equal	No	3.247
13	70	M/W/C	603,550	M no W	50	Equal	No	3.745
14	66	M/W/C	603,550	M no W	50	Equal	No	3.759
15	70	M no W	603,550	M no W	50	Equal	No	3.585
16	66	M no W	603,550	M no W	50	Equal	No	3.598
17	70	M only	603,550	M no W	50	Equal	No	3.367
18	66	M only	603,550	M no W	50	Equal	No	3.379
19	70	M/W/C	625,550	M no W	50	Equal	No	3.753
20	66	M/W/C	625,550	M no W	50	Equal	No	3.767
21	70	M no W	625,550	M no W	50	Equal	No	3.593
22	66	M no W	625,550	M no W	50	Equal	No	3.606

23	70	M only	625,550	M no W	50	Equal	No	3.375
24	66	M only	625,550	M no W	50	Equal	No	3.387
25	70	M/W/C	603,550	M only	50	Equal	No	4.013
26	66	M/W/C	603,550	M only	50	Equal	No	4.029
27	70	M no W	603,550	M only	50	Equal	No	3.835
28	66	M no W	603,550	M only	50	Equal	No	3.850
29	70	M only	603,550	M only	50	Equal	No	3.584
30	66	M only	603,550	M only	50	Equal	No	3.598
31	70	M/W/C	625,550	M only	50	Equal	No	4.023
32	66	M/W/C	625,550	M only	50	Equal	No	4.038
33	70	M no W	625,550	M only	50	Equal	No	3.844
34	66	M no W	625,550	M only	50	Equal	No	3.859
35	70	M only	625,550	M only	50	Equal	No	3.593
36	66	M only	625,550	M only	50	Equal	No	3.606
37	70	M/W/C	603,550	M/W/C	70	Equal	No	3.624
38	66	M/W/C	603,550	M/W/C	70	Equal	No	3.638
39	70	M no W	603,550	M/W/C	70	Equal	No	3.466
40	66	M no W	603,550	M/W/C	70	Equal	No	3.479
41	70	M only	603,550	M/W/C	70	Equal	No	3.257
42	66	M only	603,550	M/W/C	70	Equal	No	3.269
43	70	M/W/C	625,550	M/W/C	70	Equal	No	3.632
44	66	M/W/C	625,550	M/W/C	70	Equal	No	3.646
45	70	M no W	625,550	M/W/C	70	Equal	No	3.474
46	66	M no W	625,550	M/W/C	70	Equal	No	3.487
47	70	M only	625,550	M/W/C	70	Equal	No	3.264

48	66	M only	625,550	M/W/C	70	Equal	No	3.276
49	70	M/W/C	603,550	M no W	70	Equal	No	3.788
50	66	M/W/C	603,550	M no W	70	Equal	No	3.802
51	70	M no W	603,550	M no W	70	Equal	No	3.624
52	66	M no W	603,550	M no W	70	Equal	No	3.637
53	70	M only	603,550	M no W	70	Equal	No	3.400
54	66	M only	603,550	M no W	70	Equal	No	3.413
55	70	M/W/C	625,550	M no W	70	Equal	No	3.797
56	66	M/W/C	625,550	M no W	70	Equal	No	3.811
57	70	M no W	625,550	M no W	70	Equal	No	3.632
58	66	M no W	625,550	M no W	70	Equal	No	3.646
59	70	M only	625,550	M no W	70	Equal	No	3.408
60	66	M only	625,550	M no W	70	Equal	No	3.420
61	70	M/W/C	603,550	M only	70	Equal	No	4.065
62	66	M/W/C	603,550	M only	70	Equal	No	4.081
63	70	M no W	603,550	M only	70	Equal	No	3.882
64	66	M no W	603,550	M only	70	Equal	No	3.897
65	70	M only	603,550	M only	70	Equal	No	3.624
66	66	M only	603,550	M only	70	Equal	No	3.637
67	70	M/W/C	625,550	M only	70	Equal	No	4.075
68	66	M/W/C	625,550	M only	70	Equal	No	4.091
69	70	M no W	625,550	M only	70	Equal	No	3.891
70	66	M no W	625,550	M only	70	Equal	No	3.907
71	70	M only	625,550	M only	70	Equal	No	3.632
72	66	M only	625,550	M only	70	Equal	No	3.646

73	70	M/W/C	603,550	M/W/C	50	Unequal	No	3.735
74	66	M/W/C	603,550	M/W/C	50	Unequal	No	3.750
75	70	M no W	603,550	M/W/C	50	Unequal	No	3.564
76	66	M no W	603,550	M/W/C	50	Unequal	No	3.578
77	70	M only	603,550	M/W/C	50	Unequal	No	3.335
78	66	M only	603,550	M/W/C	50	Unequal	No	3.347
79	70	M/W/C	625,550	M/W/C	50	Unequal	No	3.744
80	66	M/W/C	625,550	M/W/C	50	Unequal	No	3.759
81	70	M no W	625,550	M/W/C	50	Unequal	No	3.572
82	66	M no W	625,550	M/W/C	50	Unequal	No	3.587
83	70	M only	625,550	M/W/C	50	Unequal	No	3.342
84	66	M only	625,550	M/W/C	50	Unequal	No	3.355
85	70	M/W/C	603,550	M no W	50	Unequal	No	3.913
86	66	M/W/C	603,550	M no W	50	Unequal	No	3.929
87	70	M no W	603,550	M no W	50	Unequal	No	3.735
88	66	M no W	603,550	M no W	50	Unequal	No	3.750
89	70	M only	603,550	M no W	50	Unequal	No	3.488
90	66	M only	603,550	M no W	50	Unequal	No	3.502
91	70	M/W/C	625,550	M no W	50	Unequal	No	3.923
92	66	M/W/C	625,550	M no W	50	Unequal	No	3.938
93	70	M no W	625,550	M no W	50	Unequal	No	3.744
94	66	M no W	625,550	M no W	50	Unequal	No	3.759
95	70	M only	625,550	M no W	50	Unequal	No	3.496
96	66	M only	625,550	M no W	50	Unequal	No	3.510
97	70	M/W/C	603,550	M only	50	Unequal	No	4.224

#							
98	66	M/W/C	603,550	M only	50	Unequal	No	4.241
99	70	M no W	603,550	M only	50	Unequal	No	4.023
100	66	M no W	603,550	M only	50	Unequal	No	4.039
101	70	M only	603,550	M only	50	Unequal	No	3.735
102	66	M only	603,550	M only	50	Unequal	No	3.750
103	70	M/W/C	625,550	M only	50	Unequal	No	4.234
104	66	M/W/C	625,550	M only	50	Unequal	No	4.252
105	70	M no W	625,550	M only	50	Unequal	No	4.033
106	66	M no W	625,550	M only	50	Unequal	No	4.050
107	70	M only	625,550	M only	50	Unequal	No	3.744
108	66	M only	625,550	M only	50	Unequal	No	3.759
109	70	M/W/C	603,550	M/W/C	70	Unequal	No	3.778
110	66	M/W/C	603,550	M/W/C	70	Unequal	No	3.794
111	70	M no W	603,550	M/W/C	70	Unequal	No	3.603
112	66	M no W	603,550	M/W/C	70	Unequal	No	3.618
113	70	M only	603,550	M/W/C	70	Unequal	No	3.366
114	66	M only	603,550	M/W/C	70	Unequal	No	3.380
115	70	M/W/C	625,550	M/W/C	70	Unequal	No	3.788
116	66	M/W/C	625,550	M/W/C	70	Unequal	No	3.803
117	70	M no W	625,550	M/W/C	70	Unequal	No	3.612
118	66	M no W	625,550	M/W/C	70	Unequal	No	3.627
119	70	M only	625,550	M/W/C	70	Unequal	No	3.375
120	66	M only	625,550	M/W/C	70	Unequal	No	3.387
121	70	M/W/C	603,550	M no W	70	Unequal	No	3.962
122	66	M/W/C	603,550	M no W	70	Unequal	No	3.978

123	70	M no W	603,550	M no W	70	Unequal	No	3.778
124	66	M no W	603,550	M no W	70	Unequal	No	3.794
125	70	M only	603,550	M no W	70	Unequal	No	3.524
126	66	M only	603,550	M no W	70	Unequal	No	3.538
127	70	M/W/C	625,550	M no W	70	Unequal	No	3.971
128	66	M/W/C	625,550	M no W	70	Unequal	No	3.987
129	70	M no W	625,550	M no W	70	Unequal	No	3.788
130	66	M no W	625,550	M no W	70	Unequal	No	3.803
131	70	M only	625,550	M no W	70	Unequal	No	3.533
132	66	M only	625,550	M no W	70	Unequal	No	3.546
133	70	M/W/C	603,550	M only	70	Unequal	No	4.282
134	66	M/W/C	603,550	M only	70	Unequal	No	4.300
135	70	M no W	603,550	M only	70	Unequal	No	4.076
136	66	M no W	603,550	M only	70	Unequal	No	4.093
137	70	M only	603,550	M only	70	Unequal	No	3.779
138	66	M only	603,550	M only	70	Unequal	No	3.794
139	70	M/W/C	625,550	M only	70	Unequal	No	4.293
140	66	M/W/C	625,550	M only	70	Unequal	No	4.311
141	70	M no W	625,550	M only	70	Unequal	No	4.086
142	66	M no W	625,550	M only	70	Unequal	No	4.104
143	70	M only	625,550	M only	70	Unequal	No	3.788
144	66	M only	625,550	M only	70	Unequal	No	3.803
145	70	M/W/C	603,550	M/W/C	50	Equal	Yes	3.868
146	66	M/W/C	603,550	M/W/C	50	Equal	Yes	3.882
147	70	M no W	603,550	M/W/C	50	Equal	Yes	3.702

148	66	M no W	603,550	M/W/C	50	Equal	Yes	3.716
149	70	M only	603,550	M/W/C	50	Equal	Yes	3.473
150	66	M only	603,550	M/W/C	50	Equal	Yes	3.485
151	70	M/W/C	625,550	M/W/C	50	Equal	Yes	3.877
152	66	M/W/C	625,550	M/W/C	50	Equal	Yes	3.891
153	70	M no W	625,550	M/W/C	50	Equal	Yes	3.710
154	66	M no W	625,550	M/W/C	50	Equal	Yes	3.724
155	70	M only	625,550	M/W/C	50	Equal	Yes	3.480
156	66	M only	625,550	M/W/C	50	Equal	Yes	3.493
157	70	M/W/C	603,550	M no W	50	Equal	Yes	4.041
158	66	M/W/C	603,550	M no W	50	Equal	Yes	4.056
159	70	M no W	603,550	M no W	50	Equal	Yes	3.868
160	66	M no W	603,550	M no W	50	Equal	Yes	3.882
161	70	M only	603,550	M no W	50	Equal	Yes	3.623
162	66	M only	603,550	M no W	50	Equal	Yes	3.636
163	70	M/W/C	625,550	M no W	50	Equal	Yes	4.050
164	66	M/W/C	625,550	M no W	50	Equal	Yes	4.065
165	70	M no W	625,550	M no W	50	Equal	Yes	3.877
166	66	M no W	625,550	M no W	50	Equal	Yes	3.891
167	70	M only	625,550	M no W	50	Equal	Yes	3.631
168	66	M only	625,550	M no W	50	Equal	Yes	3.644
169	70	M/W/C	603,550	M only	50	Equal	Yes	4.345
170	66	M/W/C	603,550	M only	50	Equal	Yes	4.362
171	70	M no W	603,550	M only	50	Equal	Yes	4.152
172	66	M no W	603,550	M only	50	Equal	Yes	4.168

173	70	M only	603,550	M only	50	Equal	Yes	3.868
174	66	M only	603,550	M only	50	Equal	Yes	3.883
175	70	M/W/C	625,550	M only	50	Equal	Yes	4.355
176	66	M/W/C	625,550	M only	50	Equal	Yes	4.372
177	70	M no W	625,550	M only	50	Equal	Yes	4.162
178	66	M no W	625,550	M only	50	Equal	Yes	4.178
179	70	M only	625,550	M only	50	Equal	Yes	3.877
180	66	M only	625,550	M only	50	Equal	Yes	3.891
181	70	M/W/C	603,550	M/W/C	70	Equal	Yes	3.910
182	66	M/W/C	603,550	M/W/C	70	Equal	Yes	3.925
183	70	M no W	603,550	M/W/C	70	Equal	Yes	3.740
184	66	M no W	603,550	M/W/C	70	Equal	Yes	3.754
185	70	M only	603,550	M/W/C	70	Equal	Yes	3.504
186	66	M only	603,550	M/W/C	70	Equal	Yes	3.517
187	70	M/W/C	625,550	M/W/C	70	Equal	Yes	3.919
188	66	M/W/C	625,550	M/W/C	70	Equal	Yes	3.934
189	70	M no W	625,550	M/W/C	70	Equal	Yes	3.749
190	66	M no W	625,550	M/W/C	70	Equal	Yes	3.763
191	70	M only	625,550	M/W/C	70	Equal	Yes	3.512
192	66	M only	625,550	M/W/C	70	Equal	Yes	3.525
193	70	M/W/C	603,550	M no W	70	Equal	Yes	4.087
194	66	M/W/C	603,550	M no W	70	Equal	Yes	4.103
195	70	M no W	603,550	M no W	70	Equal	Yes	3.910
196	66	M no W	603,550	M no W	70	Equal	Yes	3.925
197	70	M only	603,550	M no W	70	Equal	Yes	3.658

198	66	M only	603,550	M no W	70	Equal	Yes	3.671
199	70	M/W/C	625,550	M no W	70	Equal	Yes	4.097
200	66	M/W/C	625,550	M no W	70	Equal	Yes	4.112
201	70	M no W	625,550	M no W	70	Equal	Yes	3.919
202	66	M no W	625,550	M no W	70	Equal	Yes	3.934
203	70	M only	625,550	M no W	70	Equal	Yes	3.666
204	66	M only	625,550	M no W	70	Equal	Yes	3.679
205	70	M/W/C	603,550	M only	70	Equal	Yes	4.401
206	66	M/W/C	603,550	M only	70	Equal	Yes	4.419
207	70	M no W	603,550	M only	70	Equal	Yes	4.203
208	66	M no W	603,550	M only	70	Equal	Yes	4.220
209	70	M only	603,550	M only	70	Equal	Yes	3.910
210	66	M only	603,550	M only	70	Equal	Yes	3.925
211	70	M/W/C	625,550	M only	70	Equal	Yes	4.412
212	66	M/W/C	625,550	M only	70	Equal	Yes	4.429
213	70	M no W	625,550	M only	70	Equal	Yes	4.213
214	66	M no W	625,550	M only	70	Equal	Yes	4.232
215	70	M only	625,550	M only	70	Equal	Yes	3.921
216	66	M only	625,550	M only	70	Equal	Yes	3.936
217	70	M/W/C	603,550	M/W/C	50	Unequal	Yes	4.032
218	66	M/W/C	603,550	M/W/C	50	Unequal	Yes	4.048
219	70	M no W	603,550	M/W/C	50	Unequal	Yes	3.847
220	66	M no W	603,550	M/W/C	50	Unequal	Yes	3.863
221	70	M only	603,550	M/W/C	50	Unequal	Yes	3.588
222	66	M only	603,550	M/W/C	50	Unequal	Yes	3.602

223	70	M/W/C	625,550	M/W/C	50	Unequal	Yes	4.042
224	66	M/W/C	625,550	M/W/C	50	Unequal	Yes	4.058
225	70	M no W	625,550	M/W/C	50	Unequal	Yes	3.857
226	66	M no W	625,550	M/W/C	50	Unequal	Yes	3.872
227	70	M only	625,550	M/W/C	50	Unequal	Yes	3.596
228	66	M only	625,550	M/W/C	50	Unequal	Yes	3.611
229	70	M/W/C	603,550	M no W	50	Unequal	Yes	4.225
230	66	M/W/C	603,550	M no W	50	Unequal	Yes	4.242
231	70	M no W	603,550	M no W	50	Unequal	Yes	4.032
232	66	M no W	603,550	M no W	50	Unequal	Yes	4.048
233	70	M only	603,550	M no W	50	Unequal	Yes	3.754
234	66	M only	603,550	M no W	50	Unequal	Yes	3.768
235	70	M/W/C	625,550	M no W	50	Unequal	Yes	4.235
236	66	M/W/C	625,550	M no W	50	Unequal	Yes	4.252
237	70	M no W	625,550	M no W	50	Unequal	Yes	4.042
238	66	M no W	625,550	M no W	50	Unequal	Yes	4.058
239	70	M only	625,550	M no W	50	Unequal	Yes	3.763
240	66	M only	625,550	M no W	50	Unequal	Yes	3.777
241	70	M/W/C	603,550	M only	50	Unequal	Yes	4.576
242	66	M/W/C	603,550	M only	50	Unequal	Yes	4.595
243	70	M no W	603,550	M only	50	Unequal	Yes	4.359
244	66	M no W	603,550	M only	50	Unequal	Yes	4.377
245	70	M only	603,550	M only	50	Unequal	Yes	4.032
246	66	M only	603,550	M only	50	Unequal	Yes	4.048
247	70	M/W/C	625,550	M only	50	Unequal	Yes	4.588

#							
248	66	M/W/C	625,550	M only	50	Unequal	Yes	4.607
249	70	M no W	625,550	M only	50	Unequal	Yes	4.370
250	66	M no W	625,550	M only	50	Unequal	Yes	4.388
251	70	M only	625,550	M only	50	Unequal	Yes	4.042
252	66	M only	625,550	M only	50	Unequal	Yes	4.058
253	70	M/W/C	603,550	M/W/C	70	Unequal	Yes	4.079
254	66	M/W/C	603,550	M/W/C	70	Unequal	Yes	4.096
255	70	M no W	603,550	M/W/C	70	Unequal	Yes	3.890
256	66	M no W	603,550	M/W/C	70	Unequal	Yes	3.905
257	70	M only	603,550	M/W/C	70	Unequal	Yes	3.622
258	66	M only	603,550	M/W/C	70	Unequal	Yes	3.637
259	70	M/W/C	625,550	M/W/C	70	Unequal	Yes	4.089
260	66	M/W/C	625,550	M/W/C	70	Unequal	Yes	4.106
261	70	M no W	625,550	M/W/C	70	Unequal	Yes	3.899
262	66	M no W	625,550	M/W/C	70	Unequal	Yes	3.915
263	70	M only	625,550	M/W/C	70	Unequal	Yes	3.631
264	66	M only	625,550	M/W/C	70	Unequal	Yes	3.645
265	70	M/W/C	603,550	M no W	70	Unequal	Yes	4.277
266	66	M/W/C	603,550	M no W	70	Unequal	Yes	4.294
267	70	M no W	603,550	M no W	70	Unequal	Yes	4.079
268	66	M no W	603,550	M no W	70	Unequal	Yes	4.096
269	70	M only	603,550	M no W	70	Unequal	Yes	3.792
270	66	M only	603,550	M no W	70	Unequal	Yes	3.807
271	70	M/W/C	625,550	M no W	70	Unequal	Yes	4.287
272	66	M/W/C	625,550	M no W	70	Unequal	Yes	4.305

Israel Population Growth Statistical Numerical Analysis Summary:

273	70	M no W	625,550	M no W	70	Unequal	Yes	4.089
274	66	M no W	625,550	M no W	70	Unequal	Yes	4.106
275	70	M only	625,550	M no W	70	Unequal	Yes	3.801
276	66	M only	625,550	M no W	70	Unequal	Yes	3.816
277	70	M/W/C	603,550	M only	70	Unequal	Yes	4.640
278	66	M/W/C	603,550	M only	70	Unequal	Yes	4.660
279	70	M no W	603,550	M only	70	Unequal	Yes	4.417
280	66	M no W	603,550	M only	70	Unequal	Yes	4.435
281	70	M only	603,550	M only	70	Unequal	Yes	4.079
282	66	M only	603,550	M only	70	Unequal	Yes	4.096
283	70	M/W/C	625,550	M only	70	Unequal	Yes	4.652
284	66	M/W/C	625,550	M only	70	Unequal	Yes	4.672
285	70	M no W	625,550	M only	70	Unequal	Yes	4.428
286	66	M no W	625,550	M only	70	Unequal	Yes	4.447
287	70	M only	625,550	M only	70	Unequal	Yes	4.089
288	66	M only	625,550	M only	70	Unequal	Yes	4.105

Analysis Summary:

Machine design analysis requires a birth rate of 3.366 to 4.252 children per woman to obtain Israel population growth from Genesis to Exodus.

Statistical numerical analysis requires a birth rate of 3.541 to 4.167 children per woman to obtain Israel population growth from Genesis to Exodus.

The claim that the Exodus of Israel from Egypt is myth based on the singular argument that the birth rate required of children per woman during Israel stay in Egypt is so high (10, 11, 12 children per woman) is mathematically invalid.

Question: Is there evidence that countries in close geographical proximity to the lands associated with the Bible have a fertility rate approximately between 3.366 to 4.252 children per woman or approximately 3.541 to 4.167 children per woman?

Answer: The author provides total fertility rate data from the U.S. Census Bureau, International Data Base, Table 028 for countries in close geographic proximity to the lands associated with the Bible for the years 1985 to 2005.

Question: How does the author account for the apparent downward inclination of the average total fertility rates for the U.S. Census Bureau, International Data Base data for the countries listed between the years 1985 to 2005?

Answer: Artificial birth control

As the use of artificial birth control around the world increases the reader should not be surprised to see an apparent downward inclination of the average total fertility rates for the U.S. Census Bureau, International Data Base data for the countries listed between the years 1985 to 2005?

The author knows of no scientific data proving artificial birth control existed during Israel stay in Egypt; therefore, it is not unreasonable to assume the average total fertility rates for women during Israel stay in Egypt is equal to or greater than the average total fertility rates for the U.S. Census Bureau, International Data Base data for the countries listed between the years 1985 to 2005.

U. S. Census Bureau, International Data Base Table 028 - Total Fertility Rate Per Woman

Country	1985	1986	1987	1988	1989	1990	1991	1992	1993	1994	1995	1996	1997	1998	1999	2000	2001	2002	2003	2004	2005
Afghanistan	7.02	7.01	7.00	6.98	6.97	6.96	6.95	6.94	6.93	6.92	6.91	6.90	6.89	6.87	6.86	6.85	6.84	6.82	6.81	6.78	6.75
Egypt		4.52	4.40	4.28	4.17	4.05	3.93	3.83	3.73	3.63	3.60	3.58	3.55	3.53	3.39	3.24	3.17	3.10	3.02	2.95	2.88
Iran		6.20	5.80	5.50	5.30	5.30	4.90	4.30	3.80	3.40	2.90	2.50	2.30	2.10	1.97	1.89	1.83	1.79	1.76	1.74	1.73
Iraq			6.46	6.34	6.22	6.09	5.95	5.83	5.72	5.60	5.48	5.36	5.24	5.11	4.99	4.87	4.75	4.63	4.52	4.40	4.28
Israel	3.17	3.14	3.10	3.10	3.07	3.07	3.04	3.04	3.01	2.97	2.94	2.95	2.94	2.97	2.94	2.96	2.91	2.91	2.96	2.92	2.85
Jordan	6.73	6.43	6.14	5.85	5.56	5.37	5.18	4.99	4.81	4.62	4.43	4.24	4.04	3.83	3.64	3.44	3.29	3.15	3.00	2.86	2.71
Kuwait	4.36	4.07	3.75	3.65	3.46	2.52	4.08	3.34	3.49	3.53	3.65	3.86	3.64	3.51	3.39	3.26	3.20	3.14	3.08	3.03	2.97
Lebanon	3.35	3.23	3.12	3.00	2.89	2.77	2.69	2.60	2.52	2.43	2.35	2.30	2.24	2.19	2.13	2.08	2.05	2.02	1.98	1.95	1.92
Oman						6.54	6.47	6.40	6.25	6.23	6.22	6.19	6.17	6.14	6.11	6.09	6.04	5.99	5.95	5.90	5.84
Pakistan	6.94	6.93	6.92	6.77	6.62	6.47	6.32	6.17	6.01	5.86	5.71	5.56	5.41	5.24	5.07	4.90	4.75	4.60	4.44	4.29	4.14
Qatar		3.99	3.95	4.16	4.05	3.97	3.38	3.51	3.52	3.33	3.16	3.05	3.00	2.98	2.84	2.80	2.82	2.67	2.62	2.50	2.51
Saudi Arabia	6.84	6.77	6.72	6.67	6.63	6.62	6.61	5.48	5.20	4.95	4.74	4.65	4.58	4.50	4.43	4.36	4.29	4.23	4.17	4.11	4.05
Syria	7.11	6.95	6.78	6.42	6.06	5.70	5.35	4.99	4.63	4.55	4.47	4.39	4.31	4.22	4.14	4.06	3.95	3.84	3.72	3.61	3.50
Turkey	3.81	3.66	3.51	3.36	3.21	3.10	3.00	2.89	2.79	2.66	2.54	2.46	2.39	2.31	2.24	2.16	2.12	2.07	2.03	1.98	1.94
UAE	5.48	5.48	5.33	5.27	5.00	4.70	4.14	3.94	3.65	3.56	3.08	2.79	2.68	2.63	2.56	2.63	2.57	2.51	2.50	2.46	2.45
Yemen	7.79	7.77	7.75	7.74	7.72	7.70	7.68	7.66	7.57	7.43	7.37	7.31	7.24	7.18	7.11	7.05	6.97	6.90	6.82	6.78	6.67
Minimum	3.17	3.14	3.10	3.00	2.89	2.52	2.69	2.60	2.52	2.43	2.35	2.30	2.24	2.10	1.97	1.89	1.83	1.79	1.76	1.74	1.73
Maximum	7.79	7.77	7.75	7.74	7.72	7.70	7.68	7.66	7.57	7.43	7.37	7.31	7.24	7.18	7.11	7.05	6.97	6.90	6.82	6.78	6.75
Average	5.71	5.44	5.38	5.27	5.13	5.06	4.98	4.74	4.60	4.48	4.35	4.26	4.16	4.08	3.99	3.92	3.85	3.77	3.71	3.64	3.57

Question: Can the author suggest any other mathematical considerations regarding Israel population growth from Genesis to Exodus?

Answer: Test opponents theory

In the introduction the author wrote about an instructor who implied that the Exodus of Israel from Egypt had to be myth because the birth rate of children per woman during Israel stay in Egypt had to be so unreasonably high (10, 11, 12 children per woman).

The author has shown that machine design analysis requires a birth rate of 3.366 to 4.252 children per woman to obtain Israel population growth from Genesis to Exodus. The author has also shown that statistical numerical analysis requires a birth rate of 3.541 to 4.167 children per woman to obtain Israel population growth from Genesis to Exodus.

While a birth rate of 10 to 12 children per woman is not required to obtain Israel population growth from Genesis to Exodus the author suggests testing the instructor's proposal of 10 to 12 children per woman. In the field of mathematics it is advisable to test your opponents theory.

Since determining Israel ending population for a birth rate of 6, 7, 8, 9, 10, 11, and 12 children per woman after 430 years in Egypt is similar to the calculations in the machine design analysis section; therefore, the author will show summary data only.

Number of children per woman	Israel ending population after 430 years in Egypt
12.0	Over 100 million people
11.0	Over 100 million people
10.0	Over 100 million people
9.0	Over 100 million people
8.0	Over 100 million people
7.0	Over 100 million people
6.0	Over 100 million people

The author knows of no scientific data proving Israel population in Egypt ever reached 100 million people; therefore, no further consideration of the instructor's proposal is necessary.

Question: The author has shown that machine design analysis requires a birth rate of 3.366 to 4.252 children per woman to obtain Israel population growth from Genesis to Exodus. The author has also shown that statistical numerical analysis requires a birth rate of 3.541 to 4.167 children per woman to obtain Israel population growth from Genesis to Exodus. Can the author provide proof that another nation has experienced a similar population growth in approximately 430 years?

Answer: 1492

According to *The World Almanac and Book of Facts 2007*, p. 459 (World Almanac Books, ISBN # 978-0-88687-995-2) Christopher Columbus and crew sighted land Oct. 12 in present-day Bahamas. The author will use the discovery of North America and South America in 1492 (2007 – 1492 equals 515 years) to compare with Israel population growth from Genesis to Exodus.

The *World Almanac and Book of Facts 2007*, (World Almanac Books, ISBN # 978-0-88687-995-2) states the following population for the partial list of North America and South America countries.

N & S America Nation	Population
Argentina	39,921,833
Bolivia	8,989,046
Brazil	188,078,227
Canada	33,098,932
Chile	16,134,219
Colombia	43,593,035
Costa Rica	4,075,261
Cuba	11,382,820
Ecuador	13,547,510
El Salvador	6,822,378
Guatemala	12,454,747
Honduras	7,326,496
Mexico	107,449,525
Nicaragua	5,570,129
Panama	3,191,319
Paraguay	6,506,464
Peru	28,302,603
United States	298,444,215
Uruguay	3,431,932
Venezuela	25,730,435
	864,051,126

The author will use machine design analysis similar to the previous machine design analysis. The author will show limited details and summary data only.

Machine Design Analysis

Starting Population

70	

Starting Population

M/W/C		

Ending Population

864	

Ending Population

M/W/C		

Life Expectancy

	70

Fractional

	Unequal

Infant Mortality

No	

The starting population assumes the word "men" means adult men, adult women, and children (70 M/W/C); therefore, the total starting population would be 70 persons.

The ending population assumes the word "men" means adult men, adult women, and children (864 M/W/C); therefore, the total ending population would be 864,051,126 persons.

Assume a human life expectancy of 70 years. Using numerical analysis to calculate the birth rate required for a total starting population of 70 persons to increase to a total ending population of 864,051,126 persons in 515 years (2007 – 1492 equals 515 years) results in a birth rate of 5.124 children.

Question: Why is the author using the population for multiple nations when Israel is a singular nation?

Answer: The author will use the population for the United States of America.

The starting population assumes the word "men" means adult men, adult women, and children (70 M/W/C); therefore, the total starting population would be 70 persons.

The ending population assumes the word "men" means adult men, adult women, and children (298 M/W/C); therefore, the total ending population would be 298,444,215 persons.

Assume a human life expectancy of 70 years. Using numerical analysis to calculate the birth rate required for a total starting population of 70 persons to increase to a total ending population of 298,444,215 persons in 515 years (2007 – 1492 equals 515 years) results in a birth rate of 4.829 children.

Question: Did the author subtract native American population from the United States population data?

Answer: According to the *World Almanac and Book of Facts 2007*, p. 606, all American Indian and Alaska native tribal grouping alone or in any combination population is 4,119,301 persons.

The starting population assumes the word "men" means adult men, adult women, and children (70 M/W/C); therefore, the total starting population would be 70 persons.

The ending population assumes the word "men" means adult men, adult women, and children (294 M/W/C); therefore, the total ending population would be 294,324,914 persons (298,444,215 – 4,119,301 equals 294,324,914 persons).

Assume a human life expectancy of 70 years. Using numerical analysis to calculate the birth rate required for a total starting population of 70 persons to increase to a total ending population of 294,324,914 persons in 515 years (2007 – 1492 equals 515 years) results in a birth rate of 4.825 children.

Question: Did the author subtract U.S. foreign-born population from the United States population data?

Answer: According to the *World Almanac and Book of Facts 2007*, p.599, the percentage of the U.S. population that is foreign-born in 2005 is 11.7 percent (298,444,215 * 0.117 equals 34,917,973 persons).

The starting population assumes the word "men" means adult men, adult women, and children (70 M/W/C); therefore, the total starting population would be 70 persons.

The ending population assumes the word "men" means adult men, adult women, and children (259 M/W/C); therefore, the total ending population would be 259,406,941 persons (298,444,215 – 4,119,301 – 34,917,973 equals 259,406,941 persons).

Assume a human life expectancy of 70 years. Using numerical analysis to calculate the birth rate required for a total starting population of 70 persons to increase to a total ending population of 259,406,941 persons in 515 years (2007 – 1492 equals 515 years) results in a birth rate of 4.791 children.

Question: Does the *World Almanac and Book of Facts 2007* have population data from the American colonies?

Answer 1: According to the *World Almanac and Book of Facts 2007*, p.594, the estimated population of American colonies is 4,600 persons in 1630.

The starting population assumes the word "men" means adult men, adult women, and children (4600 M/W/C); therefore, the total starting population would be 4,600 persons.

The ending population assumes the word "men" means adult men, adult women, and children (259 M/W/C); therefore, the total ending population would be 259,406,941 persons.

Assume a human life expectancy of 70 years. Using numerical analysis to calculate the birth rate required for a total starting population of 4,600 persons to increase to a total ending population of 259,406,941 persons in 375 years (2005 – 1630 equals 375 years) results in a birth rate of 4.803 children.

Answer 2: According to the *World Almanac and Book of Facts 2007*, p.594, the estimated population of American colonies is 50,400 persons in 1650.

The starting population assumes the word "men" means adult men, adult women, and children (50,400 M/W/C); therefore, the total starting population would be 50,400 persons.

The ending population assumes the word "men" means adult men, adult women, and children (259 M/W/C); therefore, the total ending population would be 259,406,941 persons.

Assume a human life expectancy of 70 years. Using numerical analysis to calculate the birth rate required for a total starting population of 50,400 persons to increase to a total ending population of 259,406,941 persons in 355 years (2005 – 1650 equals 355 years) results in a birth rate of 4.147 children.

Answer 3: According to the *World Almanac and Book of Facts 2007*, p.594, the estimated population of American colonies is 111,900 persons in 1670.

The starting population assumes the word "men" means adult men, adult women, and children (111 M/W/C); therefore, the total starting population would be 111,900 persons.

The ending population assumes the word "men" means adult men, adult women, and children (259 M/W/C); therefore, the total ending population would be 259,406,941 persons.

Assume a human life expectancy of 70 years. Using numerical analysis to calculate the birth rate required for a total starting population of 111,900 persons to increase to a total ending population of 259,406,941 persons in 335 years (2005 – 1670 equals 335 years) results in a birth rate of 4.040 children.

Design	Country	Starting Population	Ending Population	Years	Birth Rate Required
001	N & S America	70	864,051,126	515	5.124
002	United States	70	298,444,215	515	4.829
003	United States	70	294,324,914	515	4.825
004	United States	70	259,406,941	515	4.791
005	United States	4,600	259,406,941	375	4.803
006	United States	50,400	259,406,941	355	4.147
007	United States	111,900	259,406,941	335	4.040

Question: Can the author give any reasons why the instructor at Sacred Heart Major Seminar erred in thinking that the birth rate of children per woman during Israel stay in Egypt had to be so high (10, 11, 12 children per woman)?

Answer 1: Numbers 14: 26-35

Numbers 14: 26-35

The Lord also said to Moses and Aaron: "How long will this wicked community grumble against me? I have heard the grumblings of the Israelites against me. Tell them: By my life, says the Lord, I will do to you just what I have heard you say. Here in the desert shall your dead bodies fall. Of all your men of twenty years or more, registered in the census, who grumbled against me, not one shall enter the land where I solemnly swore to settle you, except Caleb, son of Jephunneh, and Joshua, son of Nun.

Your little ones, however, who you said would be taken as booty, I will bring in, and they shall appreciate the land you spurned. But as for you, your bodies shall fall here in the desert, here where your children must wander for forty years, suffering for your faithlessness, till the last of you lies dead in the desert. Forty days you spent in scouting the land; forty years shall you suffer for your crimes: one year for each day. Thus you will realize what it means to oppose me. I, the Lord, have sworn to do this to all this wicked community that conspired against me: here in the desert they shall die to the last man."

The classical biblical definition of one generation is forty years. Exodus 12: 40 states that the Israelites stayed in Egypt for 430 years. In other words 10.75 generations of Israelites during Israel stay in Egypt (430 divided by 40 equals 10.75).

Can the author estimate the actual number of Israelite generations during Israel stay in Egypt? The author's statistical numerical analysis of Israel population growth from Genesis to Exodus required a birth rate of 3.541 to 4.167 children per woman. In others words of Israel population growth from Genesis to Exodus required a birth rate of 3.854 plus or minus 0.313 children per woman.

The author's statistical numerical analysis of Israel population growth from Genesis to Exodus analyzed 288 combinations, but the author showed statistical numerical analysis summary data only. The author's machine design analysis 159 showed analysis details and had a required birth rate of 3.868 children per woman for Israel population growth from Genesis to Exodus.

3.868 children per woman is equal to 1.934 children per person increase in one generation (assume one man and one woman per married couple; 3.868 divided by 2 equals 1.934). Machine design analysis 159 ending population after 430 years was 1,209,072 persons. We need to determine when Israel population was approximately 625,166 persons (1,209,072 divided by 1.934 equals 625,166 persons). Machine design analysis 159 ending population was 570,126 persons after 395 years and 634,762 persons after 400 years. Using linear approximation Israel ending population was approximately 625,166 persons at approximately 399.3 years. For machine design analysis 159 one generation is approximately 30.7 years (430 years minus 399.3 years equals 30.7 years). In other words 14.00 generations of Israelites during Israel stay in Egypt (430 divided by 30.7 equals 14.00). The difference in population growth for Israel between a birth rate of 1.934 children per person increase in a 40 year generation over 430 years to birth rate of 1.934 children per person increase in a 30.7 year generation over 430 years is 8.53 ($1.934^{14.00}$ divided by $1.934^{10.75}$ equals 8.53).

Answer 2: Matthew 1: 18-25 and Luke 1: 26-38

Matthew 1: 18-25

Now this is how the birth of Jesus Christ came about. When his mother Mary was betrothed to Joseph, but before they lived together, she was found with child through the holy Spirit. Joseph her husband, since he was a righteous man, yet unwilling to expose her to shame, decided to divorce her quietly. Such was his intention when, behold, the angel of the Lord appeared to him in a dream and said, "Joseph, son of David, do not be afraid to take Mary your wife into your home. For it is through the holy Spirit that this child has been conceived in her. She will bear a son and you are to name him Jesus, because he will save his people from their sins. All this took place to fulfill what the Lord had said through the prophet: "Behold, the virgin shall be with child and bear a son, and they shall name him Emmanuel," which means "God is with us." When Joseph awoke, he did as the angel of the Lord had commanded him and took his wife into his home. He had no relations with her until she bore a son, and he named him Jesus.

Luke 1: 26-38

In the sixth month, the angel Gabriel was sent from God to a town of Galilee called Nazareth, to a virgin betrothed to a man named Joseph, of the house of David, and the virgin's name was Mary. And coming to her, he said, "Hail, favored one! The Lord is with you." But she was greatly troubled at what was said and pondered what sort of greeting this might be. Then the angel said to her, Do not be afraid, Mary, for you have found favor with God.

Behold, you will conceive in your womb and bear a son, and you shall name him Jesus. He will be great and will be called Son of the Most High, and the Lord God will give him the throne of David his father, and he will rule over the house of Jacob forever, and of his kingdom there will be no end." But Mary said to the angel, "How can this be, since I have no relations with a man? And the angel said to her in reply, "The holy Spirit will come upon you, and the power of the Most High will overshadow you. Therefore the child to be born will be called holy, the Son of God. And behold, Elizabeth, your relative, has also conceived a son in her old age, and this is the sixth month for her who was called barren; for nothing will be impossible for God." Mary said, "Behold, I am the handmaid of the Lord. May it be done to me according to your word." Then the angel departed from her.

The Blessed Virgin Mary conceived Our Lord and Savior Jesus Christ when she was probably 12, 13, 14, or 15 years old. Since Mary was not expected to have a career in a multinational corporation there was no need for a girl in Israel to delay childbearing until her mid to late-thirties. Since Mary was not expected to obtain a bachelor degree or masters degree there was no need for a girl in Israel to delay childbearing until her mid to late-twenties. Since Mary was not expected to attend a secular public high school there was no need for a girl in Israel to delay childbearing until her late-teens. In Luke 1: 26-38 Mary said "How can this be, since I have no relations with a man?" Mary at 12, 13, 14, or 15 years old understood that it was time to be married, have sexual relations with a man, and start bearing children.

The author when analyzing Israel population growth from Genesis to Exodus unequal fractional assumes women bearing children occurs similar to U.S. census bureau data over time.

The Blessed Virgin Mary conceived Our Lord and Savior Jesus Christ when she was probably 12, 13, 14, or 15 years old. Women of Israel during their stay in Egypt probably had children sooner in life rather than later as shown by the U.S. Census Bureau data. The author will assume 9.6 % of the children are born to women between 10 to 15 years of age rather than 9.6 % of the children being born to women between 35 to 39 years of age resulting in the following author modified unequal fractional for analyzing Israel population growth from Genesis to Exodus.

Age Range of Women	U.S. Census Bureau Data	Author modified
10 to 15 years	0.0 % of the children born	9.6 % of the children born
15 to 20 years	12.5 % of the children born	12.5 % of the children born
20 to 25 years	27.0 % of the children born	27.0 % of the children born
25 to 30 years	28.6 % of the children born	28.6 % of the children born
30 to 35 years	22.2 % of the children born	22.2 % of the children born
35 to 40 years	9.6 % of the children born	0.0 % of the children born

Machine design analysis 235 has a starting population of 70 persons and an ending population of 1,252,547 persons using U.S. Census Bureau data for unequal fractional. Machine design analysis 235 has a starting population of 70 persons and an ending population of 3,070,055 persons using author modified data for unequal fractional.

						Numerical analysis 235
Birth rate per (2) persons per (25) years	4.235		Total starting population of 70 persons.			
Birth rate per (1) person per (25) years	2.118		Total ending population 1,251,100 persons			
Birth rate per (1) person per (5) years	0.424		Israel stayed in Egypt 430 years.			
Infant Mortality Rate (per 1000)	73.70					

0 to 5 Yrs	5 to 10 Yrs	10 to 15 Yrs	15 to 20 Yrs	20 to 25 Yrs	25 to 30 Yrs	30 to 35 Yrs	35 to 40 Yrs	40 to 45 Yrs	45 to 50 Yrs	Total Population	Years
			0.125	0.270	0.286	0.222	0.096				
7.00	7.00	7.00	7.00	7.00	7.00	7.00	7.00	7.00	7.00	70.0	0
13.7	7.0	7.0	7.0	7.0	7.0	7.0	7.0	7.0	7.0	76.7	5
13.7	13.7	7.0	7.0	7.0	7.0	7.0	7.0	7.0	7.0	83.4	10
13.7	13.7	13.7	7.0	7.0	7.0	7.0	7.0	7.0	7.0	90.1	15
13.7	13.7	13.7	13.7	7.0	7.0	7.0	7.0	7.0	7.0	96.9	20
15.4	13.7	13.7	13.7	13.7	7.0	7.0	7.0	7.0	7.0	105.2	25
18.9	15.4	13.7	13.7	13.7	13.7	7.0	7.0	7.0	7.0	117.1	30
22.7	18.9	15.4	13.7	13.7	13.7	13.7	7.0	7.0	7.0	132.8	35
25.6	22.7	18.9	15.4	13.7	13.7	13.7	13.7	7.0	7.0	151.4	40
27.3	25.6	22.7	18.9	15.4	13.7	13.7	13.7	13.7	7.0	171.7	45
29.0	27.3	25.6	22.7	18.9	15.4	13.7	13.7	13.7	13.7	193.8	50
32.8	29.0	27.3	25.6	22.7	18.9	15.4	13.7	13.7	13.7	212.8	55
38.2	32.8	29.0	27.3	25.6	22.7	18.9	15.4	13.7	13.7	237.3	60
44.1	38.2	32.8	29.0	27.3	25.6	22.7	18.9	15.4	13.7	267.7	65
49.4	44.1	38.2	32.8	29.0	27.3	25.6	22.7	18.9	15.4	303.3	70
54.1	49.4	44.1	38.2	32.8	29.0	27.3	25.6	22.7	18.9	342.1	75
59.7	54.1	49.4	44.1	38.2	32.8	29.0	27.3	25.6	22.7	382.8	80
67.2	59.7	54.1	49.4	44.1	38.2	32.8	29.0	27.3	25.6	427.3	85
76.6	67.2	59.7	54.1	49.4	44.1	38.2	32.8	29.0	27.3	478.3	90
87.0	76.6	67.2	59.7	54.1	49.4	44.1	38.2	32.8	29.0	538.0	95
97.4	87.0	76.6	67.2	59.7	54.1	49.4	44.1	38.2	32.8	606.4	100
108.3	97.4	87.0	76.6	67.2	59.7	54.1	49.4	44.1	38.2	681.9	105
120.7	108.3	97.4	87.0	76.6	67.2	59.7	54.1	49.4	44.1	764.4	110
135.8	120.7	108.3	97.4	87.0	76.6	67.2	59.7	54.1	49.4	856.1	115
153.4	135.8	120.7	108.3	97.4	87.0	76.6	67.2	59.7	54.1	960.1	120
172.9	153.4	135.8	120.7	108.3	97.4	87.0	76.6	67.2	59.7	1,078.9	125
193.9	172.9	153.4	135.8	120.7	108.3	97.4	87.0	76.6	67.2	1,213.1	130
216.7	193.9	172.9	153.4	135.8	120.7	108.3	97.4	87.0	76.6	1,362.6	135
242.7	216.7	193.9	172.9	153.4	135.8	120.7	108.3	97.4	87.0	1,528.7	140
272.7	242.7	216.7	193.9	172.9	153.4	135.8	120.7	108.3	97.4	1,714.5	145
307.0	272.7	242.7	216.7	193.9	172.9	153.4	135.8	120.7	108.3	1,924.1	150
345.2	307.0	272.7	242.7	216.7	193.9	172.9	153.4	135.8	120.7	2,161.0	155
387.2	345.2	307.0	272.7	242.7	216.7	193.9	172.9	153.4	135.8	2,427.5	160
434.0	387.2	345.2	307.0	272.7	242.7	216.7	193.9	172.9	153.4	2,725.7	165
486.7	434.0	387.2	345.2	307.0	272.7	242.7	216.7	193.9	172.9	3,059.1	170
546.7	486.7	434.0	387.2	345.2	307.0	272.7	242.7	216.7	193.9	3,432.9	175
614.4	546.7	486.7	434.0	387.2	345.2	307.0	272.7	242.7	216.7	3,853.4	180
690.1	614.4	546.7	486.7	434.0	387.2	345.2	307.0	272.7	242.7	4,326.8	185
774.5	690.1	614.4	546.7	486.7	434.0	387.2	345.2	307.0	272.7	4,858.6	190
868.9	774.5	690.1	614.4	546.7	486.7	434.0	387.2	345.2	307.0	5,454.7	195
975.1	868.9	774.5	690.1	614.4	546.7	486.7	434.0	387.2	345.2	6,122.8	200

1,095.0	975.1	868.9	774.5	690.1	614.4	546.7	486.7	434.0	387.2	6,872.6	205
1,229.8	1,095.0	975.1	868.9	774.5	690.1	614.4	546.7	486.7	434.0	7,715.2	210
1,380.8	1,229.8	1,095.0	975.1	868.9	774.5	690.1	614.4	546.7	486.7	8,662.0	215
1,549.9	1,380.8	1,229.8	1,095.0	975.1	868.9	774.5	690.1	614.4	546.7	9,725.2	220
1,739.5	1,549.9	1,380.8	1,229.8	1,095.0	975.1	868.9	774.5	690.1	614.4	10,917.9	225
1,952.6	1,739.5	1,549.9	1,380.8	1,229.8	1,095.0	975.1	868.9	774.5	690.1	12,256.1	230
2,192.3	1,952.6	1,739.5	1,549.9	1,380.8	1,229.8	1,095.0	975.1	868.9	774.5	13,758.2	235
2,461.5	2,192.3	1,952.6	1,739.5	1,549.9	1,380.8	1,229.8	1,095.0	975.1	868.9	15,445.2	240
2,763.6	2,461.5	2,192.3	1,952.6	1,739.5	1,549.9	1,380.8	1,229.8	1,095.0	975.1	17,339.9	245
3,102.2	2,763.6	2,461.5	2,192.3	1,952.6	1,739.5	1,549.9	1,380.8	1,229.8	1,095.0	19,467.0	250
3,482.2	3,102.2	2,763.6	2,461.5	2,192.3	1,952.6	1,739.5	1,549.9	1,380.8	1,229.8	21,854.3	255
3,909.1	3,482.2	3,102.2	2,763.6	2,461.5	2,192.3	1,952.6	1,739.5	1,549.9	1,380.8	24,533.7	260
4,388.7	3,909.1	3,482.2	3,102.2	2,763.6	2,461.5	2,192.3	1,952.6	1,739.5	1,549.9	27,541.6	265
4,927.3	4,388.7	3,909.1	3,482.2	3,102.2	2,763.6	2,461.5	2,192.3	1,952.6	1,739.5	30,919.0	270
5,531.6	4,927.3	4,388.7	3,909.1	3,482.2	3,102.2	2,763.6	2,461.5	2,192.3	1,952.6	34,711.1	275
6,209.7	5,531.6	4,927.3	4,388.7	3,909.1	3,482.2	3,102.2	2,763.6	2,461.5	2,192.3	38,968.2	280
6,970.9	6,209.7	5,531.6	4,927.3	4,388.7	3,909.1	3,482.2	3,102.2	2,763.6	2,461.5	43,746.9	285
7,825.7	6,970.9	6,209.7	5,531.6	4,927.3	4,388.7	3,909.1	3,482.2	3,102.2	2,763.6	49,111.0	290
8,785.5	7,825.7	6,970.9	6,209.7	5,531.6	4,927.3	4,388.7	3,909.1	3,482.2	3,102.2	55,133.0	295
9,863.1	8,785.5	7,825.7	6,970.9	6,209.7	5,531.6	4,927.3	4,388.7	3,909.1	3,482.2	61,893.9	300
11,072.7	9,863.1	8,785.5	7,825.7	6,970.9	6,209.7	5,531.6	4,927.3	4,388.7	3,909.1	69,484.4	305
12,430.4	11,072.7	9,863.1	8,785.5	7,825.7	6,970.9	6,209.7	5,531.6	4,927.3	4,388.7	78,005.6	310
13,954.4	12,430.4	11,072.7	9,863.1	8,785.5	7,825.7	6,970.9	6,209.7	5,531.6	4,927.3	87,571.3	315
15,665.6	13,954.4	12,430.4	11,072.7	9,863.1	8,785.5	7,825.7	6,970.9	6,209.7	5,531.6	98,309.6	320
17,586.8	15,665.6	13,954.4	12,430.4	11,072.7	9,863.1	8,785.5	7,825.7	6,970.9	6,209.7	110,364.9	325
19,743.7	17,586.8	15,665.6	13,954.4	12,430.4	11,072.7	9,863.1	8,785.5	7,825.7	6,970.9	123,898.9	330
22,164.9	19,743.7	17,586.8	15,665.6	13,954.4	12,430.4	11,072.7	9,863.1	8,785.5	7,825.7	139,092.9	335
24,882.8	22,164.9	19,743.7	17,586.8	15,665.6	13,954.4	12,430.4	11,072.7	9,863.1	8,785.5	156,150.0	340
27,934.0	24,882.8	22,164.9	19,743.7	17,586.8	15,665.6	13,954.4	12,430.4	11,072.7	9,863.1	175,298.5	345
31,359.5	27,934.0	24,882.8	22,164.9	19,743.7	17,586.8	15,665.6	13,954.4	12,430.4	11,072.7	196,794.8	350
35,205.2	31,359.5	27,934.0	24,882.8	22,164.9	19,743.7	17,586.8	15,665.6	13,954.4	12,430.4	220,927.3	355
39,522.5	35,205.2	31,359.5	27,934.0	24,882.8	22,164.9	19,743.7	17,586.8	15,665.6	13,954.4	248,019.5	360
44,369.2	39,522.5	35,205.2	31,359.5	27,934.0	24,882.8	22,164.9	19,743.7	17,586.8	15,665.6	278,434.2	365
49,810.0	44,369.2	39,522.5	35,205.2	31,359.5	27,934.0	24,882.8	22,164.9	19,743.7	17,586.8	312,578.6	370
55,918.0	49,810.0	44,369.2	39,522.5	35,205.2	31,359.5	27,934.0	24,882.8	22,164.9	19,743.7	350,909.7	375
62,775.1	55,918.0	49,810.0	44,369.2	39,522.5	35,205.2	31,359.5	27,934.0	24,882.8	22,164.9	393,941.1	380
70,473.3	62,775.1	55,918.0	49,810.0	44,369.2	39,522.5	35,205.2	31,359.5	27,934.0	24,882.8	442,249.5	385
79,115.5	70,473.3	62,775.1	55,918.0	49,810.0	44,369.2	39,522.5	35,205.2	31,359.5	27,934.0	496,482.3	390
88,817.4	79,115.5	70,473.3	62,775.1	55,918.0	49,810.0	44,369.2	39,522.5	35,205.2	31,359.5	557,365.7	395
99,708.9	88,817.4	79,115.5	70,473.3	62,775.1	55,918.0	49,810.0	44,369.2	39,522.5	35,205.2	625,715.1	400
111,936.0	99,708.9	88,817.4	79,115.5	70,473.3	62,775.1	55,918.0	49,810.0	44,369.2	39,522.5	702,445.8	405
125,662.6	111,936.0	99,708.9	88,817.4	79,115.5	70,473.3	62,775.1	55,918.0	49,810.0	44,369.2	788,585.9	410
141,072.6	125,662.6	111,936.0	99,708.9	88,817.4	79,115.5	70,473.3	62,775.1	55,918.0	49,810.0	885,289.3	415
158,372.3	141,072.6	125,662.6	111,936.0	99,708.9	88,817.4	79,115.5	70,473.3	62,775.1	55,918.0	993,851.6	420
177,793.3	158,372.3	141,072.6	125,662.6	111,936.0	99,708.9	88,817.4	79,115.5	70,473.3	62,775.1	1,115,726.9	425
199,595.8	177,793.3	158,372.3	141,072.6	125,662.6	111,936.0	99,708.9	88,817.4	79,115.5	70,473.3	1,252,547.5	430

Birth rate per (2) persons per (25) years	4.235
Birth rate per (1) person per (25) years	2.118
Birth rate per (1) person per (5) years	0.424
Infant Mortality Rate (per 1000)	73.70

Total starting population of 70 persons.
Total ending population 1,251,100 persons
Israel stayed in Egypt 430 years.

Numerical analysis 235
Author modified unequal fractional

0 to 5 Yrs	5 to 10 Yrs	10 to 15 Yrs	15 to 20 Yrs	20 to 25 Yrs	25 to 30 Yrs	30 to 35 Yrs	35 to 40 Yrs	40 to 45 Yrs	45 to 50 Yrs	Total Population	Years
		0.096	0.125	0.270	0.286	0.222					
7.00	7.00	7.00	7.00	7.00	7.00	7.00	7.00	7.00	7.00	70.0	0
13.7	7.0	7.0	7.0	7.0	7.0	7.0	7.0	7.0	7.0	76.7	5
13.7	13.7	7.0	7.0	7.0	7.0	7.0	7.0	7.0	7.0	83.4	10
13.7	13.7	13.7	7.0	7.0	7.0	7.0	7.0	7.0	7.0	90.1	15
15.0	13.7	13.7	13.7	7.0	7.0	7.0	7.0	7.0	7.0	98.1	20
16.6	15.0	13.7	13.7	13.7	7.0	7.0	7.0	7.0	7.0	107.8	25
20.2	16.6	15.0	13.7	13.7	13.7	7.0	7.0	7.0	7.0	120.9	30
24.2	20.2	16.6	15.0	13.7	13.7	13.7	7.0	7.0	7.0	138.1	35
27.7	24.2	20.2	16.6	15.0	13.7	13.7	13.7	7.0	7.0	158.9	40
29.5	27.7	24.2	20.2	16.6	15.0	13.7	13.7	13.7	7.0	181.3	45
32.7	29.5	27.7	24.2	20.2	16.6	15.0	13.7	13.7	13.7	207.0	50
37.7	32.7	29.5	27.7	24.2	20.2	16.6	15.0	13.7	13.7	231.0	55
43.7	37.7	32.7	29.5	27.7	24.2	20.2	16.6	15.0	13.7	261.0	60
50.4	43.7	37.7	32.7	29.5	27.7	24.2	20.2	16.6	15.0	297.7	65
56.8	50.4	43.7	37.7	32.7	29.5	27.7	24.2	20.2	16.6	339.6	70
63.4	56.8	50.4	43.7	37.7	32.7	29.5	27.7	24.2	20.2	386.3	75
71.4	63.4	56.8	50.4	43.7	37.7	32.7	29.5	27.7	24.2	437.5	80
81.6	71.4	63.4	56.8	50.4	43.7	37.7	32.7	29.5	27.7	494.9	85
93.5	81.6	71.4	63.4	56.8	50.4	43.7	37.7	32.7	29.5	560.7	90
106.4	93.5	81.6	71.4	63.4	56.8	50.4	43.7	37.7	32.7	637.6	95
120.3	106.4	93.5	81.6	71.4	63.4	56.8	50.4	43.7	37.7	725.2	100
135.7	120.3	106.4	93.5	81.6	71.4	63.4	56.8	50.4	43.7	823.2	105
153.8	135.7	120.3	106.4	93.5	81.6	71.4	63.4	56.8	50.4	933.3	110
175.1	153.8	135.7	120.3	106.4	93.5	81.6	71.4	63.4	56.8	1,058.0	115
199.4	175.1	153.8	135.7	120.3	106.4	93.5	81.6	71.4	63.4	1,200.5	120
226.3	199.4	175.1	153.8	135.7	120.3	106.4	93.5	81.6	71.4	1,363.5	125
256.3	226.3	199.4	175.1	153.8	135.7	120.3	106.4	93.5	81.6	1,548.5	130
290.4	256.3	226.3	199.4	175.1	153.8	135.7	120.3	106.4	93.5	1,757.3	135
329.6	290.4	256.3	226.3	199.4	175.1	153.8	135.7	120.3	106.4	1,993.4	140
374.6	329.6	290.4	256.3	226.3	199.4	175.1	153.8	135.7	120.3	2,261.5	145
425.5	374.6	329.6	290.4	256.3	226.3	199.4	175.1	153.8	135.7	2,566.7	150
482.8	425.5	374.6	329.6	290.4	256.3	226.3	199.4	175.1	153.8	2,913.9	155
547.5	482.8	425.5	374.6	329.6	290.4	256.3	226.3	199.4	175.1	3,307.5	160
621.1	547.5	482.8	425.5	374.6	329.6	290.4	256.3	226.3	199.4	3,753.5	165
705.0	621.1	547.5	482.8	425.5	374.6	329.6	290.4	256.3	226.3	4,259.1	170
800.4	705.0	621.1	547.5	482.8	425.5	374.6	329.6	290.4	256.3	4,833.2	175
908.7	800.4	705.0	621.1	547.5	482.8	425.5	374.6	329.6	290.4	5,485.5	180
1,031.1	908.7	800.4	705.0	621.1	547.5	482.8	425.5	374.6	329.6	6,226.2	185
1,169.8	1,031.1	908.7	800.4	705.0	621.1	547.5	482.8	425.5	374.6	7,066.4	190
1,327.5	1,169.8	1,031.1	908.7	800.4	705.0	621.1	547.5	482.8	425.5	8,019.3	195
1,506.7	1,327.5	1,169.8	1,031.1	908.7	800.4	705.0	621.1	547.5	482.8	9,100.6	200

1,710.3	1,506.7	1,327.5	1,169.8	1,031.1	908.7	800.4	705.0	621.1	547.5	10,328.0	205
1,941.1	1,710.3	1,506.7	1,327.5	1,169.8	1,031.1	908.7	800.4	705.0	621.1	11,721.6	210
2,202.8	1,941.1	1,710.3	1,506.7	1,327.5	1,169.8	1,031.1	908.7	800.4	705.0	13,303.4	215
2,499.7	2,202.8	1,941.1	1,710.3	1,506.7	1,327.5	1,169.8	1,031.1	908.7	800.4	15,098.1	220
2,836.9	2,499.7	2,202.8	1,941.1	1,710.3	1,506.7	1,327.5	1,169.8	1,031.1	908.7	17,134.5	225
3,219.7	2,836.9	2,499.7	2,202.8	1,941.1	1,710.3	1,506.7	1,327.5	1,169.8	1,031.1	19,445.6	230
3,654.3	3,219.7	2,836.9	2,499.7	2,202.8	1,941.1	1,710.3	1,506.7	1,327.5	1,169.8	22,068.8	235
4,147.3	3,654.3	3,219.7	2,836.9	2,499.7	2,202.8	1,941.1	1,710.3	1,506.7	1,327.5	25,046.2	240
4,706.6	4,147.3	3,654.3	3,219.7	2,836.9	2,499.7	2,202.8	1,941.1	1,710.3	1,506.7	28,425.3	245
5,341.3	4,706.6	4,147.3	3,654.3	3,219.7	2,836.9	2,499.7	2,202.8	1,941.1	1,710.3	32,259.9	250
6,061.9	5,341.3	4,706.6	4,147.3	3,654.3	3,219.7	2,836.9	2,499.7	2,202.8	1,941.1	36,611.5	255
6,879.8	6,061.9	5,341.3	4,706.6	4,147.3	3,654.3	3,219.7	2,836.9	2,499.7	2,202.8	41,550.2	260
7,808.0	6,879.8	6,061.9	5,341.3	4,706.6	4,147.3	3,654.3	3,219.7	2,836.9	2,499.7	47,155.4	265
8,861.3	7,808.0	6,879.8	6,061.9	5,341.3	4,706.6	4,147.3	3,654.3	3,219.7	2,836.9	53,516.9	270
10,056.5	8,861.3	7,808.0	6,879.8	6,061.9	5,341.3	4,706.6	4,147.3	3,654.3	3,219.7	60,736.6	275
11,413.1	10,056.5	8,861.3	7,808.0	6,879.8	6,061.9	5,341.3	4,706.6	4,147.3	3,654.3	68,930.0	280
12,952.8	11,413.1	10,056.5	8,861.3	7,808.0	6,879.8	6,061.9	5,341.3	4,706.6	4,147.3	78,228.5	285
14,700.2	12,952.8	11,413.1	10,056.5	8,861.3	7,808.0	6,879.8	6,061.9	5,341.3	4,706.6	88,781.4	290
16,683.4	14,700.2	12,952.8	11,413.1	10,056.5	8,861.3	7,808.0	6,879.8	6,061.9	5,341.3	100,758.2	295
18,933.9	16,683.4	14,700.2	12,952.8	11,413.1	10,056.5	8,861.3	7,808.0	6,879.8	6,061.9	114,350.8	300
21,488.0	18,933.9	16,683.4	14,700.2	12,952.8	11,413.1	10,056.5	8,861.3	7,808.0	6,879.8	129,776.9	305
24,386.7	21,488.0	18,933.9	16,683.4	14,700.2	12,952.8	11,413.1	10,056.5	8,861.3	7,808.0	147,283.9	310
27,676.6	24,386.7	21,488.0	18,933.9	16,683.4	14,700.2	12,952.8	11,413.1	10,056.5	8,861.3	167,152.4	315
31,410.2	27,676.6	24,386.7	21,488.0	18,933.9	16,683.4	14,700.2	12,952.8	11,413.1	10,056.5	189,701.4	320
35,647.6	31,410.2	27,676.6	24,386.7	21,488.0	18,933.9	16,683.4	14,700.2	12,952.8	11,413.1	215,292.4	325
40,456.4	35,647.6	31,410.2	27,676.6	24,386.7	21,488.0	18,933.9	16,683.4	14,700.2	12,952.8	244,335.7	330
45,913.9	40,456.4	35,647.6	31,410.2	27,676.6	24,386.7	21,488.0	18,933.9	16,683.4	14,700.2	277,296.9	335
52,107.7	45,913.9	40,456.4	35,647.6	31,410.2	27,676.6	24,386.7	21,488.0	18,933.9	16,683.4	314,704.4	340
59,137.2	52,107.7	45,913.9	40,456.4	35,647.6	31,410.2	27,676.6	24,386.7	21,488.0	18,933.9	357,158.2	345
67,114.9	59,137.2	52,107.7	45,913.9	40,456.4	35,647.6	31,410.2	27,676.6	24,386.7	21,488.0	405,339.2	350
76,168.8	67,114.9	59,137.2	52,107.7	45,913.9	40,456.4	35,647.6	31,410.2	27,676.6	24,386.7	460,019.9	355
86,444.0	76,168.8	67,114.9	59,137.2	52,107.7	45,913.9	40,456.4	35,647.6	31,410.2	27,676.6	522,077.2	360
98,105.3	86,444.0	76,168.8	67,114.9	59,137.2	52,107.7	45,913.9	40,456.4	35,647.6	31,410.2	592,505.9	365
111,339.8	98,105.3	86,444.0	76,168.8	67,114.9	59,137.2	52,107.7	45,913.9	40,456.4	35,647.6	672,435.4	370
126,359.7	111,339.8	98,105.3	86,444.0	76,168.8	67,114.9	59,137.2	52,107.7	45,913.9	40,456.4	763,147.5	375
143,405.8	126,359.7	111,339.8	98,105.3	86,444.0	76,168.8	67,114.9	59,137.2	52,107.7	45,913.9	866,096.9	380
162,751.3	143,405.8	126,359.7	111,339.8	98,105.3	86,444.0	76,168.8	67,114.9	59,137.2	52,107.7	982,934.3	385
184,706.6	162,751.3	143,405.8	126,359.7	111,339.8	98,105.3	86,444.0	76,168.8	67,114.9	59,137.2	1,115,533.2	390
209,623.7	184,706.6	162,751.3	143,405.8	126,359.7	111,339.8	98,105.3	86,444.0	76,168.8	67,114.9	1,266,019.7	395
237,902.1	209,623.7	184,706.6	162,751.3	143,405.8	126,359.7	111,339.8	98,105.3	86,444.0	76,168.8	1,436,806.9	400
269,995.4	237,902.1	209,623.7	184,706.6	162,751.3	143,405.8	126,359.7	111,339.8	98,105.3	86,444.0	1,630,633.5	405
306,418.1	269,995.4	237,902.1	209,623.7	184,706.6	162,751.3	143,405.8	126,359.7	111,339.8	98,105.3	1,850,607.6	410
347,754.1	306,418.1	269,995.4	237,902.1	209,623.7	184,706.6	162,751.3	143,405.8	126,359.7	111,339.8	2,100,256.5	415
394,666.5	347,754.1	306,418.1	269,995.4	237,902.1	209,623.7	184,706.6	162,751.3	143,405.8	126,359.7	2,383,583.2	420
447,907.4	394,666.5	347,754.1	306,418.1	269,995.4	237,902.1	209,623.7	184,706.6	162,751.3	143,405.8	2,705,130.9	425
508,330.5	447,907.4	394,666.5	347,754.1	306,418.1	269,995.4	237,902.1	209,623.7	184,706.6	162,751.3	3,070,055.6	430

Author's Comments: Luke 1: 31-38

Behold, you will conceive in your womb and bear a son, and you shall name him Jesus. He will be great and will be called Son of the Most High, and the Lord God will give him the throne of David his father, and he will rule over the house of Jacob forever, and of his kingdom there will be no end." But Mary said to the angel, "How can this be, since I have no relations with a man? And the angel said to her in reply, "The holy Spirit will come upon you, and the power of the Most High will overshadow you. Therefore the child to be born will be called holy, the Son of God. And behold, Elizabeth, your relative, has also conceived a son in her old age, and this is the sixth month for her who was called barren; for nothing will be impossible for God." Mary said, "Behold, I am the handmaid of the Lord. May it be done to me according to your word." Then the angel departed from her.

The author was born in 1961 and raised in Southwest Detroit, Michigan, USA. In the author's formative years it was not uncommon to see families with 3, 4, 5, or 6 children. Examples from the author's childhood friends: the Ke___ had four boys (M, M, C, and M) and the Ma___ family had three boys and one girl (B, G, D, and D). Luke 1: 31-38 speaks of Elizabeth, who was called barren, conceiving in her old age. It was not that long ago having zero or one child was thought to be a curse and having 3, 4, 5, or 6 children was thought to be a blessing. Women of Israel during their stay in Egypt probably had children sooner in life rather than later. Women of Israel during their stay in Egypt probably wanted to have 3, 4, 5, or 6 children rather than zero, 1, or 2 children. It is only recently with the invention of artificial birth control and the propagation of the birth control mentality in secular and Christian thought that women started to delay having children and started to have smaller families.

Machine design analysis requires a birth rate of 3.366 to 4.252 children per woman to obtain Israel population growth from Genesis to Exodus.

Statistical numerical analysis requires a birth rate of 3.541 to 4.167 children per woman to obtain Israel population growth from Genesis to Exodus.

The claim that the Exodus of Israel from Egypt is myth based on the singular argument that the birth rate required of children per woman during Israel stay in Egypt is so high (10, 11, 12 children per woman) is mathematically invalid.

www.ingramcontent.com/pod-product-compliance
Lightning Source LLC
Chambersburg PA
CBHW020245290526
45784CB00003B/1104